THE CARBON WAVE

THE MIT PRESS'S PUBLISHING MISSION BENEFITS FROM THE GENEROSITY OF OUR DONORS, INCLUDING DIANA CHAPMAN WALSH.

THE CARBON WAVE

A STORY OF DEMOCRACY, PARENTHOOD, AND THE RACE TO PROTECT OUR PLANET

Leah C. Stokes

THE MIT PRESS
CAMBRIDGE, MASSACHUSETTS
LONDON, ENGLAND

The MIT Press
Massachusetts Institute of Technology
77 Massachusetts Avenue
Cambridge, MA 02139
mitpress.mit.edu

The MIT Press would like to thank the anonymous peer reviewers who provided comments on drafts of this book. The generous work of academic experts is essential for establishing the authority and quality of our publications. We acknowledge with gratitude the contributions of these otherwise uncredited readers.

This book was set in ITC Stone and Futura Std by New Best-set Typesetters Ltd. Printed and bound in the United States of America.

Library of Congress Cataloging-in-Publication Data is available.

ISBN: 978-0-262-05477-5

10 9 8 7 6 5 4 3 2 1

EU Authorised Representative: Easy Access System Europe, Mustamäe tee 50, 10621 Tallinn, Estonia | Email: gpsr.requests@easproject.com

For my daughters, and for all living beings.
May the fossil fuel era end.

Something is wrong, I know it, if I don't keep my attention on eternity.
May I be the tiniest nail in the house of the universe, tiny but useful.
—Mary Oliver

CONTENTS

CAST OF CHARACTERS

The following individuals, listed in order of appearance, are featured throughout the book. With few exceptions, political staff and activists are referred to by their first names, while elected and appointed politicians are referred to by their last names. Below are the roles held at the time the story takes place.

CONGRESS

Adrian Deveny, Director of Energy and Environmental Policy for Senate Majority Leader Chuck Schumer

Gerry Petrella, Policy Director for Senate Majority Leader Chuck Schumer

Tim Ryder, Legislative Assistant for Senate Majority Leader Chuck Schumer

Keane Bhatt, Policy Director for the Congressional Progressive Caucus

Anna Taylor, Tax Counsel for Senate Majority Leader Chuck Schumer

WHITE HOUSE AND ADMINISTRATION

Sonia Aggarwal, Special Assistant to the President for Climate Policy, Innovation, and Deployment

Brian Deese, Director of the White House National Economic Council

Jahi Wise, Special Assistant to the President for Climate and Finance

Gina McCarthy, inaugural White House National Climate Advisor

Ron Klain, White House Chief of Staff

ADVOCATES

Sam Ricketts, Co-Founder of Evergreen Action and Senior Fellow at the Center for American Progress

Varshini Prakash, Executive Director of the Sunrise Movement

Julian NoiseCat, Vice President of Policy and Strategy at Data for Progress

Bill McKibben, author and founder of 350.org and Third Act

Christy Goldfuss, Senior Vice President for Energy and Environment Policy at the Center for American Progress

John Podesta, Founder of the Center for American Progress, after serving in the Clinton and Obama administrations

Hasan Nazar, Director of Policy and Business Development at Tesla

INTRODUCTION

The first time my newborn's eyes saw the sky, it was a dull gray. The first time her lungs breathed outdoor air, it was filled with smoke. On the morning she came home from the hospital, a wildfire burned just miles up the coast from Santa Barbara.

It was October 14, 2021, the day before her due date, and 75 long nights after her birth. She'd spent that time inside a NICU incubator. But that evening my baby slept in her bassinet for the first time, reunited with her twin sister, who'd come home a few weeks earlier.

The next morning, our family woke up under one roof. I'd anticipated this day for the past nine months. To mark the occasion, we were going to sing, eat chocolate cake, and celebrate.

The twins were not all I'd been growing over that long year. I was having a strange, high-risk pregnancy: gestating policy ideas and babies. During this time, I worked with a team of fellow dreamers to develop an ambitious plan to rapidly grow clean electricity and slash fossil fuels. What began as a seed—a line in a platform for a presidential candidate—grew into a policy paper, then a network of advocates, and finally a bill introduced in Congress. Along the way, we'd even managed to get President Joe Biden to commit to our plan. Now it was part of a climate package, the biggest in history, slowly winding its way through Congress.

Both the pregnancy and the policy proved precarious. There were weeks when it looked like my future children might die, and the policy seemed destined to pass. And other weeks when I was sure the babies would live, and the policy would crumble. I desperately wanted everything to work out, but often that felt impossible.

That jarring feeling hit me again on October 15, 2021, my due date. Just as we were about to sit down for our first dinner as a family, I checked my email. When I saw a message from Coral Davenport, my heart sank. As one of the longest-serving climate journalists at *The New York Times*, Coral was often the first to get the scoop. She wanted to talk as soon as possible. I immediately knew why.

Ten days earlier, as my second baby was growing healthy enough to finally come home, I began to hear that the clean energy policy I'd poured so much into was dying. Rumors of its demise grew daily. On one of our last nights at the hospital, my friend Sonia Aggarwal, who worked at the White House, called. It was dinnertime, and I needed to drive to the hospital before visiting hours ended, but I also knew I needed to pick up the phone.

As one of my closest friends and allies in this work, Sonia said she wanted to be the one to tell me that the policy's prognosis was not good. She would keep pushing, and so should I, but the case increasingly seemed terminal. Senator Joe Manchin from West Virginia had hardened against the idea. He held the pivotal fiftieth vote in the Senate and could veto anything he opposed. A man who'd made a fortune in coal could not support a rapid transition away from dirty fossil fuels. That night, surrounded by the incessant beeping and alarms of the NICU, I held my tiny daughter and wept.

One week later, with both of my children now home, I picked up the phone and Coral's voice came over the line: "I've got the story. I know the policy is dead." Once published, those words would be true. She wanted a quote, to explain why this mattered in simple terms, but I couldn't find the words. "Can I call you back in five minutes?" was all I could manage.

The larger climate package was still on the table, but the clean electricity plan, the heart of the bill, was on the chopping block. Its loss would blow a giant hole through the bill. Without it, I feared we wouldn't be able to cut pollution fast enough. My first instinct was despair—but despair is contagious. Plus, there were many other policies left in the package, policies worth saving.

With just a small window of time, I reached out to three colleagues I'd worked with over the past nine months. Each one held a different view, leaving me feeling like I was running through the stages of grief in rapid succession. One person was stuck in denial: "It's not dead yet. Do not confirm it. Keep pushing." Another wanted to bargain: "Maybe we can still fix this. There must be something else we can do." The third counseled acceptance: "It's gone. We need to move on and focus on everything we can still save." I knew my friend Adrian Deveny, who was leading the Senate negotiations for Majority Leader Chuck Schumer, wanted me to move on too. He had been saying for weeks that we were likely to lose this policy.

With these conflicting voices in my head, I tried to compose a brief comment. Coral had the story right; she'd heard it from multiple sources. There was no point in denying or trying to change the facts. I also knew the loss was too big to accept quietly. Instead, I would try to communicate the gravity of the situation while leaving space for hope. I wrote down, word for word, what I intended to say and practiced it aloud to my husband, who was sitting on our bed watching.

When I called Coral back, I did my best to stick to my lines. But I found myself saying more than I meant to. Emotion came spilling out as my voice trailed off, "It's pretty sad . . . it's so sad."

My husband, noticing I'd gone off script, signaled that I needed to stop talking. When I hung up, I reassured him, "Oh, the part about me being sad? Don't worry, she won't use that."

That night, we still sat down for dinner. We still sang. We still ate cake. I did my best to hold the grief and the joy simultaneously. I'd found myself in the middle of my life, in the middle of a policy fight, in the middle of the climate crisis. In my brief 35 years, polluters had pumped out more than half of all carbon emissions in human history. Even on that happy day, the crisis was still there, casting a shadow. It had already made the air too smoke-filled to take my newborns outside.

The next morning, I walked to the corner store to pick up a copy of *The New York Times* and found the article on the front page above the fold: "Crucial Element of Climate Plan Likely to Be Cut: Manchin Blocks Effort." It opened with a sober reading of the facts. "The most powerful part of President Biden's climate agenda—a program to rapidly replace the nation's coal- and gas-fired power plants with wind, solar and nuclear energy—will likely be dropped from the massive budget bill pending in

Congress, according to congressional staffers and lobbyists familiar with the matter."

Manchin's staff all but confirmed it. The White House wouldn't comment. And then, halfway down the page, were my words: "This is absolutely the most important climate policy in the package. We fundamentally need it to meet our climate goals. That's just the reality. And now we can't. So this is pretty sad."

They say you should choose your words carefully. That you shouldn't say something you wouldn't want to read on the cover of a newspaper. Because words can have consequences you can't predict. In the coming days, young people would begin a hunger strike outside the White House, staffers would find new ways to cut pollution, and pressure would grow on Congress to act. But that was all in the future. That Saturday, all I felt was the overwhelming feeling that I had failed—failed the world and, perhaps most acutely, failed my children.

* * *

I first heard about climate change two decades earlier, in 2001. At the time, I was in high school and my geography teacher—a brilliant British man with thick white hair—would draw fluffy clouds on the chalkboard to show how mountains create rain shadows and deserts. He was lighthearted and kind, and encouraged students to speak up. Plus, he had a funny ritual of closing each lecture by saying: "From the bottom of my heart, and the depths of my conviction: Get lost." I took every class he taught. But one morning he brought up global warming, saying it was an unproven theory. Just because temperatures were rising alongside carbon dioxide did not mean humans were to blame. Correlation was not causation, after all. Despite my admiration for him, this lesson struck me as misguided. I don't know why, but I knew he was wrong.

In my four years of college at the University of Toronto, I became immersed in climate change through coursework and extracurriculars. One evening in the winter of 2007, a few months before graduation, I walked toward the university's Convocation Hall. The round building was imposing, with thirty-foot-tall columns that sheltered huge wooden doors. When I had classes in this building, I quickly learned that if I showed up late for a lecture, I'd be stuck in the balcony, because 1,700 other people had arrived before me. But that night I didn't have to worry about where to sit. There was a reserved spot in the front row.

Surrounded by fellow climate activists, I watched Al Gore give his famous slideshow. In Toronto, he was like a rockstar—when ticket sales for the event went live, the website crashed. After the talk ended, my friends and I were invited to attend a reception. I even got to shake Al Gore's hand.

Gore's documentary, *An Inconvenient Truth*, had come out nine months earlier, creating widespread public awareness about climate change for the first time. In it, he presents the same slideshow I saw live. Just four days after this talk, it would win best documentary at the Oscars.

When I think back on that film two decades later, strangely, the thing I remember most vividly is the credits. After all the charts and facts came the typical black screen. But in addition to listing the people who made the film, a series of messages appeared: "The climate crisis can be solved. You can reduce your carbon emissions. In fact, you can even reduce your carbon emissions to zero." These words appeared before me like a sacred text. And then the answer came: "Buy efficient appliances and light bulbs . . . Recycle."

The idea that individuals could solve climate change on their own was ubiquitous at the time. It was 2007, and "carbon footprints" were all the rage. The claim was that we were each responsible for producing some amount of emissions, and we each needed to work, relentlessly, to reduce them. It felt like everyone I knew was calculating how much pollution their "lifestyle" created. As a psychology major, I was susceptible to this argument. In college, I ran campaigns to get students to change their behavior and save energy. Then my first job after graduation was conducting a carbon audit for a nonprofit. Once I figured out exactly how much pollution they made, the proposed solution was buying carbon offsets. What else could we do? No one seemed to know. It was counting for counting's sake.

During this same period, I joined 50 million people across the globe and sat in the dark for an hour. "Earth Hour" was a global campaign aiming to show the world's commitment to climate action. In Toronto, the iconic CN Tower switched off its lights, alongside hundreds of other buildings, dropping energy use 9 percent in an instant. But as that dark hour passed, I began to wonder: how would turning off the lights solve the climate crisis?

When I think back now on that film's closing credits, I can see they simply weren't true. No individual can reduce their carbon emissions to zero—no amount of personal sacrifice will do it. Until we get rid of fossil fuel infrastructure in every corner of our society, individual behavior changes are mere nibbles around the edges of the problem.

Of course, this was not Al Gore's fault—as usual, fossil fuel executives were to blame. In 2004, the fossil fuel company BP put a carbon footprint calculator on their website encouraging people to estimate how much their behaviors, like food choices and commutes, affected the planet. For the next three years, they spent upward of $100 million a year promoting the idea. The phrase "carbon footprint" was initially unknown, but by the time *An Inconvenient Truth* hit theaters in 2006, it was everywhere. No wonder it became the film's framework for action.

It similarly took me a decade to understand why my brilliant high school teacher had spread climate denial. Beginning in 1998, the American Petroleum Institute funded online curricula that promoted fossil fuels and cast doubt on climate science. These dirty industries were doing everything they could to delay policy action and protect their profits. The longer I worked on climate change, the more I realized there were villains in this story. They were pushing us to obsess over our personal carbon footprints to keep us separated and distracted.

The carbon footprint paradigm didn't just shame people for driving to work—it also discouraged having children. Academic articles were published claiming the number-one way to reduce your impact was forgoing a child. According to this perspective, the average American is responsible for 16 tons of carbon pollution each year—including every baby. If you don't have that kid, theoretically the world makes less pollution.

My whole life, I struggled with whether to have children. Not because I worried they would add to the problems of this world. And only slightly because of the fires that burned hotter and closer each year. Instead, I feared that motherhood would blur my focus, drain my energy, and reduce my impact.

Typically, when people ask if it's moral to have a child with climate change accelerating, they are asking two things at once. They want to know if it's okay to bring a child into a doomed world. And they want to know if the kid would only make the crisis worse. To me, both of these questions are fundamentally flawed: we are not doomed, and everyday people—let alone newborn babies—are not responsible for our dirty energy system.

We have too easily accepted the idea that we're all to blame for the climate crisis. Whether or not a baby is born, our world will keep churning out pollution. Everyone alive today was born into a society that runs on fossil fuels. That's even true for my grandparents, born almost a century ago. But it doesn't have to be this way. We've jumped from climate denial to climate despair without pausing to consider a third option: actually doing something about the problem. Our dirty energy system is not inevitable.

This is the work, regardless of whether or not someone chooses to have a child. Because there are myriad other beings worthy of our care. Our nieces and nephews. Our neighbors. The migratory birds whose call we recognize when they return home each year, sounding like an old friend. We cannot divorce ourselves from these beings' fate whether or not we become parents.

When we start thinking this way, the challenge shifts from minimizing our impact on the planet to maximizing it. At first, this is hard to do, because it's a muscle we haven't used in a while. But we can teach ourselves to imagine again.

Let's replace the carbon footprint with a new idea: the carbon wave. We are carbon-based beings. After oxygen, it's the most common element in our bodies. And we are full of energy. We can make change—especially when we join with others. Rather than thinking of a single person leaving a footprint in the sand, think of waves in the ocean. These waves are made by strong, sustained winds blowing in the same direction. Over generations, these waves are built, as thousands of people work together to create change, adding to the effort of the people before them. And just as waves wash away footprints on a beach, our carbon waves can erase our carbon footprints. Whatever negative impacts we have on the planet from living in a fossil fuel society can be offset by our activism, if we join with others to change the system. Working together, we can make waves.

It's the same idea as a group of strangers at a ballgame, rising together to send a wave around the stadium. After they do this, they feel a little closer. If we see ourselves as isolated and small, we will remain that way. But if we see ourselves as connected, we become more powerful. Together, we can make the biggest wave possible to rip fossil fuels out of our society.

Of course, waves are less controllable than footprints. The waves we make alongside others will have less certain impacts. But their potential is orders of magnitude larger. Working together, we can grow these waves and watch them change history.

This book is a story of a group of people who worked through our democracy to create one big carbon wave—passing the largest climate law in American history. It begins in 2018, when young activists started pushing for a "Green New Deal." The next year, more than two dozen Democrats ran for the presidential nomination amid the public's growing attention to climate. By the time President Joe Biden won the 2020 election, climate action was firmly on the agenda. Over the next 20 months, Congress and the White House worked on a bill, as activists pushed them relentlessly. In August 2022, President Biden finally signed the Inflation Reduction Act into law, with north of $400 billion in climate investments.

The truth is, thousands of people created this carbon wave—some working for decades to envision a federal climate law. We built on the momentum they created, growing the wave even bigger. There were dozens of people in the White House, hundreds of congressional staffers, and endless climate and environmental justice organizations. There were journalists and activists and civil servants. As much as I want to tell all their stories, no single book could chronicle all of this work.

Instead, this book follows a few dozen people who worked closely for years, forming one part of the much larger carbon wave. Adrian Deveny started as a Senate staffer in 2010 with the goal of passing a climate law. He joined Majority Leader Chuck Schumer's team a decade later, becoming his chief negotiator on climate. Sonia Aggarwal honed her technical expertise in nonprofits, before joining the Biden administration in its first days. She worked on the bill with her colleagues in the White House climate office. Sam Ricketts started as a junior staffer in Congress in 2009, where he watched a landmark climate bill pass the House of Representatives, only to flame out in the Senate. That failure compelled him to spend the next decade trying to finish the job. Throughout the first two years of the Biden administration, Ricketts and I worked in partnership with hundreds of advocates pushing for this giant bill. To braid together our stories, the narrative moves between the Senate, the White House, and outside of government, jumping back and forth in time.

As we lived this story, the stakes felt high. We were fighting for the planet and everyone who lived on it. Then Adrian, Sonia, Sam, and I had children, and that feeling only intensified. As we stared down the possibility that Congress might waste another decade, we worried about the broken world we could be leaving our kids.

The main characters in this book are not politicians. They are staffers, activists, and scientists. In telling their personal stories alongside their work, I aim to show that flawed, busy people can change the world. This book is also a love letter to democracy, even when it, too, is flawed. It's our democracy that creates space for this activism, allowing us to join with others to change society. By unpacking the blow-by-blows of how this climate law was made, I hope to demystify the process and encourage more people to get involved in shaping local, state, and even federal laws. Everyone can help create carbon waves that move us away from fossil fuels.

For more than three years, this massive climate law rippled out across society. When this story began in 2018, fossil fuels made up the majority of new electricity being added to the grid. After the law passed, 96 percent of new additions were clean energy. It also spurred a renaissance in clean manufacturing, with $170 billion of private investment pouring into making solar panels, electric vehicles, batteries, and other clean technologies in the United States. Millions of American families used the law's incentives to stick solar on their roof. In 2022, heat pumps outsold gas furnaces for the first time, and over four million were installed in 2024. This was real progress, and a lot of it could not be easily reversed.

After Donald Trump won the 2024 presidential election, it was clear the climate law and our democracy were under attack. Within days of his inauguration in January 2025, the executive orders targeting clean energy began. In a repeat of his first term, Trump started to hollow out agencies—though this time, he did it on steroids, firing thousands of federal workers and incentivizing tens of thousands to quit. With Republicans controlling both the House and Senate, Congress quickly focused on repealing the climate law. Yet strangely, even though not a single Republican voted for it in 2022, over a dozen rallied to defend parts of it from repeal. The final package gutted most of Biden's climate investments, but not all of them. When Trump signed his bill into law on July 4, 2025, it was a dark day.

Even as I watched so much of this progress under attack, I remained hopeful. During the three years the law was in full effect, solar costs fell dramatically. Plus, Republicans had left federal support for batteries in place: given a few more years, they would similarly fall in cost. With artificial intelligence and electrification driving new electricity demand, and gas plants facing years-long backlogs for equipment, I knew solar and batteries would continue to get built. If people kept pushing, clean energy would be inevitable.

Progress doesn't move in a straight line, much as we might hope, and even as the sand in the hourglass runs short. But this story is not just a work of history: it's a reminder that we can and must join together to tackle the climate crisis. The choices we make now will determine our future. The majority of the American public wants climate action—they have for decades. We must keep pushing our governments, corporations, and communities to act. We can break the cycle and refuse to pass our dirty energy system to yet another generation.

Climate change is not our destiny. Together we can change the future.

ACT I

DEFEAT

1

THE GREEN NEW DEAL

Summer 2008 was an electric time. Barack Obama was running for president, and my new American friends were spending their weekends canvassing in swing states. I had just moved from Toronto to New York City to start a master's program at Columbia University. After finishing my undergraduate degree in psychology, I was disillusioned with individual behavior change and carbon footprints. I realized that changing the energy system required changing public policy. The only problem was, I had no idea how that worked. Columbia's program was designed to train students for careers in government. Over the next year, I took courses in everything from climatology to toxicology and started to understand how laws were made.

One afternoon, in the small tree-lined park that flanked my dorm, I saw activists selling political buttons. I looked down at the table, my eyes glancing at Shepard Fairey's iconic image of Obama in red, white, and blue, the word HOPE emblazoned across the bottom. But I chose a different button to stick on my backpack, one that showed Obama alongside his running mate, Joe Biden. Something about Biden's frankness appealed to me. Even when he messed up or said the wrong thing, he came across as honest and kind. I liked the guy.

That November, I crammed into the student lounge with hundreds of classmates to watch the election results roll in on a tiny television screen stuck near the ceiling. The hours passed quickly, with one state after another going to the Democrats—first Pennsylvania, then Ohio, then New Mexico. By 11:00 p.m., even Fox News had called the race. Nearby Harlem erupted in joy, as people flooded into the streets. I stayed up late, celebrating like everyone else. There was so much hope that Obama would deliver: on healthcare, yes, but on climate change too.

Initially, progress was swift. I can remember where I was standing on June 26, 2009, when I heard the House of Representatives had passed the Waxman-Markey bill. It was the farthest a major climate package had ever made it in Congress. A law was now just one Senate vote away.

But that never happened. Two months later Senator Ted Kennedy died, causing the Democrats to lose their sixty-vote supermajority. The next year, West Virginia Governor Joe Manchin decided to run for the Senate. In his campaign ad, he shot that climate bill with a gun.

The issue slipped from the congressional agenda and the public's attention for the next decade. But I stayed focused on it, starting a doctorate in public policy at the Massachusetts Institute of Technology in 2010. With each passing year, I learned to memorize the latest levels of carbon dioxide in the atmosphere: first 388, then 400, then 415. The number just kept rising. And still politicians in Washington failed to act.

In 2015, I graduated and took my first job as a professor at the University of California, Santa Barbara. The campus had a long history in the environmental movement. In early 1969, when an oil spill hit the city, the community mobilized, creating several organizations that helped spur policy change in DC. The first Earth Day was held the very next year, and the university set up an environmental studies department. Almost 50 years later, the campus remained focused on these issues.

As a Canadian who'd spent my entire life in places where winter came fast and hard, it was mesmerizing to watch the palm trees' long limbs bending in the wind. Moving to California was even more exciting because the state was leading the world on climate action. In those first few years teaching, I learned how to talk about the problem to a larger audience in lecture halls filled with students. And I continued to research and write. But two years into my time as a professor, a fire began burning near my home.

In December 2017, the academic quarter was just wrapping up when strong winds hit Southern California. The longest drought in recorded history created a tinderbox. When winds drew two power lines together, sparks flew, hit the dry ground, and ignited a fire. The wildfire traveled

12 miles within hours. Five days later, the air was toxic, and we needed to leave. For weeks, we watched as the fire burned closer and closer to our home, coming within a few miles before it was finally extinguished. At the time, the Thomas Fire was the largest wildfire in California's modern history.

We had only been home for a few days when I woke in the middle of the night to pounding rain. Half an inch of rain fell in five minutes, a torrent so fast that the scorched hillsides could not absorb it. Instead, the land slid away, causing mudslides. Helicopters flew back and forth above my home as rescuers frantically tried to save lives.

The drought, the fires, the heavy rain: all of this was predicted in climate models. Why weren't more people saying this was a climate disaster? How many more times would people need to hear adjectives in their superlative form before they spotted a pattern? Largest, rainiest, driest, deadliest. I wanted to help connect what was happening to climate change, so I stayed up late to write an article. It ran two days later on January 11, 2018, in *The New York Times* under the headline "Climate Change in My Backyard." This was my first big piece of public writing.

My hope was that some of the people on the front line of this climate disaster would become the champions we needed. At the time, I was thinking of my famous neighbors—people like Oprah Winfrey, who filmed her backyard swollen with mud, and Ellen DeGeneres, who posted a video next to a rushing creek. What I didn't realize at the time was that the fire had changed me.

A few months earlier, in September 2017, a still-unknown Alexandria Ocasio-Cortez watched as Hurricane Maria devastated Puerto Rico. Electricity went down for months. Like thousands of Puerto Ricans, her grandfather died in the storm's aftermath. She could see how climate change was hitting her community harder than others.

And it wasn't just Ocasio-Cortez's community: the year before, she'd traveled to Standing Rock, North Dakota, to see firsthand how the fossil fuel industry was harming Indigenous people. It was mere weeks after the 2016 election, and the resistance to Trump had found its center in a makeshift camp along the Missouri River. People had united to oppose an oil pipeline.

Serendipitously, the day after Ocasio-Cortez left the protest, a progressive organization reached out to ask her to run for Congress. Moved by the activists at Standing Rock, she said yes. In 2018, she ran a long-shot primary campaign in a Bronx district occupied by a ten-term congressman in House leadership. Her platform included an idea that some had bandied about before, but no one had developed in earnest: a Green New Deal. As she told journalists that summer, "The Green New Deal we are proposing will be similar in scale to the mobilization efforts seen in World War II or the Marshall Plan. We must again invest in the development, manufacturing, deployment, and distribution of energy, but this time green energy." In June, she won the primary. Her election signaled the beginning of a broader shift: climate change was back on the political agenda.

That fall, the world's leading climate science organization, the Intergovernmental Panel on Climate Change—often called the IPCC—released a short report concluding that limiting warming to 1.5°C would require emissions to fall by almost half by 2030. The news ricocheted around the world, with headlines like "We Have 12 Years to Limit Climate Catastrophe." In my own neighborhood, someone printed a copy and stuck it in the free lending library, labeling it "important reading." The report changed how many people thought about the problem: it wasn't something we could deal with in 50 years; it was something we had to deal with today.

In the November 2018 midterm elections, two years into the first Trump administration, Democrats won big, paving the way for Nancy Pelosi to return as Speaker of the House after eight years as the minority leader. Days later, a group of young activists from the Sunrise Movement staged a sit-in at her office. Their bright yellow-and-black signs read "We have 12 years. What's your plan?" referencing that scientific report. Newly elected Representative Ocasio-Cortez joined the action that day. She told the organizers they needed a clear ask and encouraged them to call for a Green New Deal. The idea broke through, sparking widespread coverage across the country. It was a new, hopeful vision for climate policy. These young people were not just talking about the climate disasters we needed to avoid—they were explaining how investments in clean energy could create good jobs. They were showing us the better world we could build together. If we only tried.

I watched these developments unfold in DC from my office by the Pacific Ocean and wanted to be a part of it. I wouldn't have to wait long. The Green New Deal held a force like gravity, pulling people into its orbit. By chance, I met some of the people tied to Ocasio-Cortez at a conference and was invited to an early gathering to develop the idea into more detailed policy. By then, she'd released a framework for the plan with Senator Ed Markey of Massachusetts.

The meeting was held in March 2019, at a hip hotel in DC's Adams Morgan neighborhood, miles away from the Capitol. A wide swath of people were invited: youth activists, environmental justice leaders, political organizers, think tank wonks, and a few academics like me. It was a contentious meeting that opened with a protest, revealing fissures in the movement that would eventually grow into large cracks. Some people believed that frontline communities facing the greatest pollution burdens were not being adequately consulted or invited to participate in the event. At the time, I didn't understand these dynamics between the climate and environmental justice movements. I was just a political science professor trying to understand how I could make my work more practical and help get a climate law passed if we got the chance.

It was at that event that I met Varshini Prakash, the executive director of the Sunrise Movement. I'd seen her photo many times, standing in the middle of Pelosi's office with her dark red jacket over a black-and-yellow T-shirt, surrounded by young people sitting on the floor. Four months had passed since the sit-in, and Varshini had ricocheted to the center of the climate movement. But she was also a recent college grad in her mid-twenties, feeling a heavy burden from the climate crisis and the work involved in fighting it. At one point during the meeting, she spoke frankly about the stress that was on her and other young activists and cried. I found her vulnerability moving.

The fact was, Varshini had already changed history. She'd given me, and millions of others, a sense of hope and possibility. It was her actions, alongside all the other organizers in the Sunrise Movement, that brought us together in that first DC convening, and in countless other meetings to come. She inspired me to believe that I could do something bigger if I tried.

Too often, the way we think about cause and effect is simplistic. We're fed a linear model that favors the immediate. One person is responsible for their own carbon footprint. One person only has one vote. One person can only do so much. From this perspective, standing against more than a century of fossil fuels running through the heart of our society, it can feel like there's nothing we could do to shake the system loose. We're too small. Mere drops. But drops in the ocean can make waves.

After the trip, my academic work felt inadequate—a feeling that had been growing for years. I didn't want to write more esoteric books on electricity policy that a few thousand people might read. By then, I'd published enough academic papers to know that securing my job long term through tenure wouldn't be a problem. As a professor, I had the freedom to use some of my time for more applied, collaborative work. That's what I wanted to do: develop practical plans to solve the crisis. But I didn't know how to begin. I would sit alone in my office on campus and ask myself questions: was it better to use a gas furnace or an electric heat pump, even if electricity running that heat pump was fueled with gas? How fast could the electricity system be cleaned up? What policies could actually pass in a closely divided Congress? These were practical research questions, but I didn't have other people to work on them with, or anything to do once I found an answer. I started to think that if I connected with activists, my work could travel a lot farther.

In an attempt to find other people interested in answering these questions, I read policy reports. In the process, I found a new polling firm called Data for Progress. Serendipitously, they reached out to talk about collaborating, and I met Julian NoiseCat, who was working there. He'd come from the climate nonprofit 350.org and was a smart organizer. On our very first call, we dreamed up climate commitments we could try to get the Democrats to adopt at their convention the next year, in 2020. The presidential primary was just beginning—although 20 candidates had already entered the race. Julian thought we could influence the politicians through polling and media work.

Before long, I was working with Julian and a motley crew of other activists to try to influence the primary. In the process, I began to fixate on electrification. After running some calculations, it became clear that electric machines were already cleaner in every state, even with fossil fuels remaining in the electricity mix. If we could ramp up heat pump and electric vehicle adoption, that would cut a lot of pollution. As I was working these ideas out, I continued to meet with new people across the climate movement. On one call, I shared my latest calculations with a scientist, but he interjected that I was still thinking too narrowly. He told me that half of heavy industry could also be electrified—food and beverage plants, pulp and paper mills, and chemicals. And then there were the diesel trains. So much could be electrified. Something inside me clicked.

I pulled up a simple pie chart that showed the United States' annual carbon pollution, sector by sector. I knew this emissions inventory well—I'd seen it over a hundred times. But this time, I came to it with fresh eyes. Rather than seeing it as a description of the problem, I looked for the solutions hidden in each wedge of the pie.

The electricity sector was simple—if you can call transforming the largest machine humans have ever built easy. We just had to replace every coal and gas plant with clean power, like wind, solar, batteries, and other technologies that didn't create carbon pollution. I'd written my first book on the laws 30 states had already passed to require more clean power. That approach was working and could be scaled nationally. Once all the electricity was clean, that would take care of around one-quarter of the climate problem.

I was less familiar with the other sectors. But now I could see that electricity was not just a mess to clean up—it was also the solution. Clean power could run almost everything we used to move around: bicycles, scooters, electric cars, school buses, postal trucks, trains, and more. If we used clean electricity to power our transportation system, it would eliminate another quarter of the country's pollution.

And then there were the buildings we lived in, worked in, played in, prayed in. These buildings did not need to use gas furnaces, let alone old oil and propane heaters. We could heat and cool them with heat pumps that ran on clean power and eliminate another eighth of our pollution.

Now I'd learned I was still thinking too small. That scientist was pointing me to yet another sector where electric machines could be deployed. I had assumed the industrial sector would be hard to decarbonize: full of steel plants that required super-high heat and cement factories that made pollution as a side effect to chemical reactions. But those kinds of problems only made up around half of the industrial sector's pollution. The other half came from systems that only required moderate heat, like a tea manufacturer drying herbs or a milk pasteurization plant. These processes could be electrified. If we did this, another eighth of the country's pollution would be gone.

A framework began to crystallize in my mind. Three-quarters of the solution came from two strategies: making clean electricity, and using electrification to run every machine we could on that clean power. It took me almost 15 years of working on climate change to figure out this simple formula. But once I found it, I saw how powerful it was. It was a simple prescription that would help people focus on the changes needed in their own homes, driveways, communities, cities, and states. Rather than fixating on small behavior changes like turning off the lights, it turned people's attention toward larger infrastructure changes, like installing a heat pump. We needed to get rid of all the fossil fuel–powered machines. If it ran on oil, coal, or gas, it had to go.

This solution also had the political benefit of splitting the two biggest opponents of climate action: fossil fuel companies and electric utilities. Both industries had promoted climate denial for decades, opposing even the smallest actions. But what if we needed more electricity to power our cars and homes and even parts of heavy industry? In that case, the utilities could grab a larger share of the energy market. They could be big winners in the transition—if they would only look five or ten years into the future, instead of staring back at the aging coal plants stuck on their balance sheets. And if we broke apart the fossil fuel companies from the utilities, we might be able to win. Rather than having to fight two powerful opponents, we could get one eight-hundred-pound gorilla to help us take down the other. At least that was my latest theory.

But for now, it was just that: a theory. To try to transform these ideas into a bill in Congress, I would need to call on some old friends—including one who was working for Senate Majority Leader Chuck Schumer.

2

RIGHT PLACE, RIGHT TIME

After nearly a decade working in Congress, Adrian Deveny was used to failure. Since he'd started working for Senator Jeff Merkley from Oregon in 2010, he'd watched just one in 20 bills become law. Before working on the Hill, he was an optimistic person—but this job had made him a pessimist. "You get trained in a fucked-up way as a Hill staffer to work on bills that you know are futile, 99 percent of the time," Adrian often said. "That's the normal state of existence." This reality bred a kind of obsessiveness that could be called paranoia—except it was warranted. Adrian would take the worst-case scenario and game out all the ways it could happen. Then he'd take the next worst-case and think that through. Rumination was a job requirement, and Adrian excelled at it.

There were rare occasions when all this worrying paid off and a bill became a law. Those times were memorialized: staffers actually hung things on the wall when they succeeded. In his home office, Adrian displayed a framed copy of a bill he'd helped pass, signed by President Obama in 2014. For three years, he'd worked on the Crooked River conflict in Oregon. It was a classic battle in the Western water wars. People kept telling Adrian: "Whiskey is for drinking, water is for fighting." Still, Senator Merkley's office was trying to overcome the conflict and meet everyone's needs—the farmers and the Tribes and the cities and the boaters and, yes, even the fish.

Trained in economics, Adrian believed people were more likely to trust a number. So he built a model in Excel. He sent the first draft over to his boss Merkley, who overhauled the whole thing on a flight back to DC—very unusual behavior for a senator. Together, they iterated until it grew sophisticated, showing all the inflows and outflows of the vast river system. Using this model, people could look at the proposed compromise and figure out if it would serve their needs. And it worked: the negotiation came together, and the bill became a law. That experience stayed with Adrian. If he ever got the chance to work on a big climate package, he knew he'd have to build another model.

Adrian grew up in a rural county in northern Maryland, one of the most conservative parts of the state. As a kid in the 1980s, he saw KKK rallies in town. Grown men dressed in white robes, hoods hiding their faces, handing out leaflets. Decades later, he could still vividly recall the day they held a cross burning in the field just across the road from his parents' home. His mom was an immigrant from Argentina, and both his parents were Spanish-language professors at local colleges, so he grew up speaking Spanish at home. In high school, he didn't talk about his background, but some of the other kids figured out he was Latino and started calling him "spic," "Chicano," even "Pocahontas." The worst was when they started fights with Adrian because of his background. At least he had good friends who would jump into these brawls to defend him. Still, heroin ripped through his high school, and by the time he graduated, two of his friends had died from overdoses. No wonder then he chose to get as far away as possible, starting college in 2000 at UC Santa Cruz. Maybe if he worked hard enough, he could outrun his problems.

I first met Adrian in 2009. He was tall and lanky, with a neatly trimmed beard. His haircut—shaved tight on the sides and left long on the top—could come across as professional or hipster, depending on the clothing he wore. As a chameleon, that looked suited him: he was smart and dedicated, yet relaxed in his off hours. At least back then. We'd both just finished master's degrees and were spending the summer in DC, working at the think tank Resources for the Future. The pay was bad, but the work was fascinating. We were surrounded by fellow nerds who enjoyed chipping away at Excel spreadsheets for weeks.

It was an uncharacteristically hopeful time for climate policy. Obama was just six months into his first term when the House of Representatives passed the Waxman-Markey bill. Climate coverage in the press was at an all-time high. The United Nations' next climate convention in Copenhagen was hotly anticipated. Adrian and I were just beginning our careers and naive enough to think that maybe our focus on climate change would be short-lived. Perhaps we would sort out this whole problem quickly and move on to world peace.

A decade passed. Adrian spent it working in the Senate, and I spent it in academia. Our careers were advancing, and we were both now married. But while we'd grown older, federal climate policy remained stagnant. The Waxman-Markey bill never made it through the Senate, and neither did other bills. When the Trump administration came to power in 2017, it eroded hundreds of environmental rules. And the Supreme Court threw out even more.

And then, in March 2018, Adrian became a father. Looking at his newborn daughter shifted something inside him. Before then, his work on climate change was abstract. Now the fight was personal. "I felt it profoundly," Adrian reflected years later. "It was about my child's future." But his optimism had vanished with the Obama administration. He stopped believing that progress was inevitable. He worried about the erosion of democracy, the rising temperatures, and the world his daughter would grow up in.

Adrian tried to control the things he could. He volunteered for local political campaigns, donated to fundraisers, and encouraged his friends to vote. He put solar panels on his roof, an electric vehicle in his garage, and an induction stove in his kitchen. The farther backward federal policy slid, the more tightly he clung to those extra actions he could take. He bought from the farmers market, but only if he could confirm the produce was organic. He recycled and rinsed out the containers first. He drank tap water, but only if it was thoroughly filtered. These daily practices masked a much larger ambition. He was just one person, but Adrian believed he could shift the course of history—if he was lucky enough to find himself in the right place at the right time.

That summer, Senator Chuck Schumer's policy director was looking for someone to take over the climate file. Gerry Petrella had boundless energy. Hailing from Long Island, he had a baritone voice and a can-do smile. Every morning, he woke up and asked himself, "How the fuck am I going to get something done today?" Gerry proved effective at getting things done, rising quickly through the ranks in Schumer's office.

He'd been carrying the climate file since Hurricane Sandy hit New York in 2012, flooding the subways and destroying homes across the region. That experience left him dedicated to the issue. But even Gerry had too much work to do. And with a newborn son he was raising with his wife, who also worked long hours for Senator Schumer, there wasn't enough time. He could not keep carrying the climate portfolio on top of all his other responsibilities. If the Democrats were serious about tackling the problem, the minority leader's office would need someone working on it full-time.

So Gerry started looking for someone. Someone with experience. Someone smart. Someone who was an operator. Years before, when Harry Reid was the majority leader, his chief of staff told Gerry: "There are too many whiners in the Democratic Party. We need more killers." The phrase stuck with him. That's what he needed.

Adrian wasn't exactly a killer—but he never whined. Gerry had gotten to know him when their offices were across from one another in the Hart Senate Building. On his door, Adrian displayed a banner that said "Keep It in the Ground"—the campaign slogan activists were using to resist fossil fuel extraction at the time. Gerry liked to tease him about the sign, asking him how exactly he planned to keep the lights on without fossil fuels. They'd become friends, and over time Gerry saw that Adrian was smart, hard-working, and pragmatic. He was well-known and well-liked across Capitol Hill. Adrian could build trust with people and solve problems. So Gerry asked him to interview for the job.

Adrian knew that working for Senator Schumer would put him in the driver's seat. He could help set the agenda and make sure climate was a priority if Democrats controlled both Congress and the White House. Still, he approached the interview with skepticism. He had zero interest in joining an office that wasn't committed to climate policy. But as he started talking with Gerry, their conversation quickly turned to strategy. Together, they imagined how they could jam a bill through the Senate with just 50 votes. Like a first date where one person leaves envisioning marriage, Gerry knew he'd found his guy. He could see that Adrian already acted like a Schumer staffer. He wasn't just out there dreaming, with pie-in-the-sky ideals—he was focused on getting shit done. And Adrian left thinking that at least Gerry got it. If other folks in the office were as committed to climate as Gerry, it would be the ideal job.

Despite the electric interview, Adrian heard nothing. Weeks passed, then months. And still nothing. After a while, he figured he hadn't been chosen. And then, in early November 2018—just days after the Green New Deal sit-in with Ocasio-Cortez at Pelosi's office—Schumer's staff called Adrian back. There was a new urgency to the issue, and Schumer didn't want to be seen as out of step—especially since that bright young congresswoman was from New York.

That winter, Adrian finally went for an interview with Schumer himself. He arrived at the office early, dressed in a sharp suit and tie, and was told to wait. The meeting was scheduled for 8:00 p.m., but Schumer was in the middle of negotiating a government spending package and was running late. Waiting on a small couch in the reception area, Adrian scrolled through his phone, trying to distract himself. Every half hour or so, Schumer would appear with Gerry trailing behind him, and they would run out the door to yet another meeting. Then a while later, they came running back in, Gerry giving Adrian a thumbs-up. As the hours passed, Adrian watched as yet another senator showed up, with her staff trailing behind, only to disappear into Schumer's office. This little reception area was like the busiest intersection in New York City: the traffic was endless, and everyone was frantic.

It was close to 10:30 p.m. before he was finally called into the meeting. The senator was in his late sixties and looked tired from the long day, but he sprang to life during the interview, throwing out rapid-fire questions. He asked Adrian to place specific politicians on a 100-point scale from most conservative to most progressive. Current senators, past senators, former presidents—all in quick succession. For Schumer, the exercise revealed two critical facts: whether a potential staffer had even heard of these people, and whether they were politically astute enough to perceive fine differences in ideology. Schumer demanded precision in the answers. When Adrian said that Bill Clinton was an 80, Schumer interrupted, "Wrong, 75."

"But he's vegan now," Adrian quipped. This made Schumer chuckle. Perhaps the kid had chutzpah.

To be an effective leader of the party, Schumer believed he had to be a 75—smack dab in the middle of the Democrats. He needed to be able to hold the trust of the centrists, like Senator Manchin, and the progressives, like Senator Markey. If he were seen as too close to one wing, he couldn't lead the whole caucus.

Shortly after his interview with Schumer, Adrian was offered the job. He took it.

When Adrian started in February 2019, it was a bit of a culture shock. He'd spent eight years in an office full of Oregonians, a polite and civil crew. Sure, there were disagreements, but they manifested as passive-aggressive tension rather than outright conflict. In contrast, Schumer's team was filled with New Yorkers: people who walked fast, talked fast, and told you exactly what they thought.

He was particularly surprised to see that Schumer's staff would sometimes yell at their boss. The senator seemed invigorated when this happened, as if he were a child again, arguing at the dinner table with his family. Schumer expected disagreements to be aired outright. He wanted people to speak up and only trusted staff who pushed back. He was more likely to get mad if his staff sat there like potted plants. "The first time Schumer yelled at me was when I was agreeing with him," Adrian told me. "He said, 'Adrian, you have to be honest with me!' And in that case, I actually did agree with him, but he didn't believe it." Adrian quickly fit in with the office culture. When he was growing up, his family debated ideas every night at the dinner table too.

If Merkley's office could be idealistic, Schumer's was ruthlessly political. He was the leader for a reason: he knew how to chart the winds and how to react to them. Schumer stayed attentive to polls, watching to see what issues mattered. To succeed in politics long-term, he knew he had to focus on the politics. And he'd been wildly successful—Schumer was nearing his fiftieth year of elected office.

The risk was that the substance could get lost if the politics overtook the policy. Adrian was aware this might happen but thought that if he worked hard enough, he could help his new boss stay focused on climate. And the truth was, all his dreams were theoretical right then. First the Democrats would need to control the Senate, making Schumer the majority leader. And then Trump would need to lose the 2020 election.

3

CAMPAIGN SEASON

In April 2019, I took a midday break from my academic work to read a short policy document. Former Representative Beto O'Rourke from Texas was running for president, and he'd just published the first climate plan of the Democratic primary. It was full of ideas that I knew well, but that most people did not: targets, tax credits, technologies. I wrote a quick evaluation, posted it on Twitter, and moved on with my day. It seemed like a small, inconsequential act. But at the time, Twitter was a platform that could amplify activists' voices and create political change.

A few days later Governor Jay Inslee of Washington released his first climate plan. Again, I wrote a quick summary and put it online. Soon I was getting called up by journalists, who asked me to compare the candidates' platforms. Not long after that, campaigns started to reach out to ask me for advice. And that's how I began to work on policy, rather than just study it.

I wasn't alone in trying to increase focus on climate change during the primary. Varshini's friends in the Sunrise Movement were showing up at campaign events to ask candidates tough questions. Journalists were organizing a climate debate. In some polls, the public was reporting it was the top issue. Politicians couldn't keep ignoring the climate crisis. Increasingly, people demanded to know what exactly they would do about it—and when.

But climate change wasn't the only thing I cared about in the primary. I was particularly keen to see a woman become president. By then, there had been 45 presidents over 243 years, and none of them were women. When Senator Kamala Harris launched her campaign in January 2019, I watched the rally live and found her speech moving. A month later,

Senator Elizabeth Warren officially entered the race. As a fellow professor, I liked her style. When I lived in Massachusetts in 2012, I'd volunteered for her first Senate race. With friends in my grad program, I sat in the middle of MIT's Infinite Corridor for weeks, pens and voter registration forms in hand, and signed people up to vote. In the end, we registered more students than any other campus. When Professor Warren became Senator Warren that fall, I went to the celebration party.

Years later, staff working on the Warren and Harris presidential campaigns reached out, asking for advice. I found myself working on a bunch of their climate plans. It was thrilling: they both had a shot at ending up in the White House. And I believed they would prioritize the issue.

Other climate advocates favored Senator Bernie Sanders from Vermont. Among progressives, conflict was brewing between Warren and Sanders supporters over who was the better candidate. I got caught up in some of this drama, because I did not agree with Bernie's proposals to shut down all the nuclear plants. But, ultimately, conflicts like these were noise. The fact was Sanders, Warren, and Harris all cared deeply about climate action. If any of them won the nomination, I knew they would listen to the climate and environmental justice movements and act boldly.

And then there was former Vice President Joe Biden. Early in the primary, he fared poorly, struggling in the Iowa caucuses and ranking low in polls. He wasn't exactly popular with climate activists either, who were skeptical that he would prioritize the issue. When his campaign released his first climate plan that June, it was particularly weak. For one thing, it included verbatim text from a fossil fuel industry group. Although Biden no doubt had nothing to do with that poor copy-paste job, journalists linked it to a plagiarism accusation that had dogged him during his first presidential run in 1988. Decades before, he'd given a speech at the Iowa State Fair, but his words mirrored an address by the UK Labour Party leader. That scandal led Biden to withdraw from the race.

The bigger problem wasn't plagiarism. It was that Biden's climate plan lacked ambition, with timelines way out to 2050. I wrote a savage review on Twitter and gave equally harsh quotes to the press. The Sunrise Movement, which was grading candidates, gave Biden's plan a failing grade—an F-minus, to be precise.

These critiques chafed at Biden. He saw himself as a climate guy—one of the earliest climate guys, in fact. He was always talking about how he introduced the first climate bill in Congress, way back in 1987. Plus, he was extremely close with his grandchildren and cared about their future. When young people in the Sunrise Movement confronted him on the campaign trail, he seemed frustrated by their perceptions.

Eventually, 29 people would enter the 2020 Democratic race, the largest in history. To call it a crowded field would be too generous—it was impossible to keep all the candidates straight. Governor Inslee stood out because climate was the reason he was running. He'd championed the issue his entire career, first as a representative in Congress and later as governor of Washington State. It was a long-shot bid for both the candidate and the climate crisis. But his presence had an unexpected effect. When his campaign published their first climate plan, an arms race of ambition began.

By August, Inslee had dropped out of the race, but not before releasing six plans. I pored through each document, finding an inspiring vision. One idea in particular caught my attention: 100 percent clean electricity by 2035. Initially, the timeline seemed ludicrous. But the idea became an earworm, repeating in my mind. I started running calculations. How much faster would clean energy need to grow? What policy could drive that kind of progress? I decided I needed to meet the person responsible for these dense documents. It turned out to be a guy named Sam Ricketts.

In October 2019, on a trip to DC, Sam Ricketts and I made plans to meet in person. I walked the long blocks from my hotel to Logan Circle, arriving early at the noisy coffee shop. I was nervous. We'd spoken on the phone once that summer, and it was awkward—largely because I'd spent the call critiquing his policy ideas. But when he walked into the coffee shop, we hit it off immediately. He was just a few inches taller than me, with a mop of messy blond hair, boyish features, and the energy of a kid. He was built for campaigning—the kind of guy who said he'd slept like three hours in the past month, and you knew he was only slightly exaggerating. Sitting across from each other at a small table, we talked nonstop, an endless stream of ideas and excitement.

We had a lot in common, including our age—it turned out we were born just one week apart. We also shared a political instinct. When it was clear that Inslee would drop out of the race in August, I called up Warren's policy director to tell him they should adopt a bunch of Inslee's ideas. I learned later that Ricketts had done the same thing—and even convinced the Warren team to hire one of Inslee's campaign staff. It was as if we'd been living in a duplex, and Ricketts was just on the other side of the wall, cooking the same things for dinner and reading the same books.

By the time we met, Ricketts had worked in politics for over a decade. He seemed to know everyone in DC, and they all called him Ricketts—there were too many Sams, and besides, it suited him. The name itself felt energetic. His phone was a Rolodex: House and Senate staffers, agency bureaucrats, lobbyists and advocates.

Born in Seattle, he'd fallen in love with DC during a college internship in the summer of 2007. He'd found living below the Mason-Dixon line thrilling. Even the stifling humidity was exciting—like the city's political activity was generating heat. After that summer, Ricketts decided to graduate early to work on a progressive political campaign with the Working Families Party. He put in months of exhausting days and helped flip a New York State Senate seat. He was hooked: the adrenaline, the stakes, the long hours. It all suited him.

Like many recent grads, Ricketts returned home to figure out what was next. He loved being back in his old stomping grounds, but Seattle wasn't exactly the political center of the country. He was living in his childhood bedroom, surrounded by his sports gear, posters of the Mariners, and books on Greek mythology. He needed to find a job. At the end of the summer, he saw an ad for a role in the district office of a congressman from Seattle—a guy named Jay Inslee. But after interviewing Ricketts, they chose someone else. He figured that was for the best and bought a one-way ticket to DC.

Once in town, he printed out a thick stack of résumés and walked around the Hill, stopping in office after office to drop off a copy. Strangely, Congressman Inslee's chief of staff called and asked Ricketts to come in for an interview. He needed someone to work on Inslee's reelection and noticed Ricketts had campaign experience. Just one week after moving to the nation's capital, he was flying back to Seattle to work for Inslee. He was right back where he'd started, but this time with a job.

Ricketts met his new boss on a conference call. Looking ahead at the calendar, Congressman Inslee was double-booked, as usual, and his most senior advisors were arguing over which event he should attend. After ten minutes of debate, Inslee cut in: "Sam Ricketts is here now. It's his job to make the decision." The line fell silent. Everyone was waiting for a twenty-two-year-old kid, on his first day on the job, to speak up. But Ricketts was never short of opinions. When he told the congressman he should go to the event rather than doing the press interview, that was the end of the conversation.

For the next six weeks, Ricketts spent every day with his new boss, driving him around the state. They formed a tight bond. A decade before, Inslee had run for governor and failed by a wide margin. But despite that loss, he wanted to try again in a few years. It was Ricketts's job to help him prepare, maintaining a packed schedule to meet as many voters as possible.

Ricketts loved being on the campaign trail, moving fast and making decisions quickly. No task was beneath or above him. His job was to do whatever needed to be done, every single hour of the day for weeks on end. Most people would find that pace exhausting, but it only revved him up. As a kid, he had so much energy his dad used to sit on him to get him to fall asleep.

The campaign was a whirlwind—and then it was over. That November 2008, Inslee and so many other Democrats won reelection in a landslide, and Obama and Biden were headed to the White House. But the election also meant that Ricketts was out of a job. Once again, he bought a one-way ticket across the country, moving from one Washington to the other and starting his job search from scratch.

Within a few weeks, Ricketts found an open position with the Democratic Congressional Campaign Committee. It was another political job, working to elect Democrats to the House. Knowing it would be competitive, he called up his old boss to ask if he would put in a good word for him. He was shocked when Inslee said no. Ricketts, confused, demanded an explanation. "I have other ideas for your future," Inslee said. He wanted Ricketts to work for him, running a new environmental caucus in the House.

When Ricketts started this new job in January 2009, the Obama administration was just beginning, and Democrats held the majority in both the House and the Senate. He worked at a manic pace, learning the details of one obscure policy after another: carbon trading, renewable portfolio standards, clean energy tax credits, home retrofit rebates, job training programs. It wasn't like being pushed into the deep end of the pool—he'd been dropped into the Mariana Trench in the middle of the Pacific Ocean. But Ricketts could swim.

Within the first month, Congress negotiated and passed a massive stimulus package that included $90 billion in funding for clean energy. By March, the House Energy and Commerce Committee had started work on Representatives Henry Waxman and Ed Markey's climate bill. By May, that bill was voted out of committee, and the next month, Speaker Nancy Pelosi passed it on the House floor. The next morning, Ricketts walked to the corner store to pick up a copy of *The New York Times*. The climate bill made the front cover. He felt like he was making history.

Months passed, and the Senate never voted. And then the 2010 midterm elections came and washed away any hope for the bill. It was the worst election for Congressional Democrats in more than 70 years. Ricketts watched as one House member after another lost their seat.

The Waxman-Markey bill centered on a price on carbon, an approach that had dominated climate policymaking for decades. The idea was simple: force polluters to pay for their carbon, and cleaner, lower-polluting options would thrive. It made sense on paper. But politically, it was terrible. Had that law passed, it would have acted as a national tax, jacking up the cost of everything from a tank of gas to monthly electricity bills. Years later, Ricketts would conclude that this was the wrong way to go about solving the problem. But at that moment, he was just sad and frustrated.

Shortly after the election, Ricketts traveled back home to Seattle to visit his parents. He was surprised when they sat him down in the living room. They had some bad news to share: his mother had just been diagnosed with a rare disease called amyloidosis, and the prognosis was not good.

In so many ways, Ricketts could trace his passion for politics back to his mother. She made it clear to all three of her children that they had to care about other people. And she believed that poetry would build empathy.

Every day during poetry month, she sent her kids to school with a poem in their lunchboxes. As a kid, Ricketts found it embarrassing to pull out a scrap of paper with lines from a sonnet in front of his friends. But when he grew up, he came to love the form, quoting Shakespeare in the middle of work meetings. "Hell is empty and all the devils are here."

Ricketts's mother took him with her when she went to vote at the polling place two blocks from their home. One year, when he was just nine years old, Ricketts accompanied her for a special election to fund the Mariners' new baseball stadium. A passionate fan, he ditched his mom in the booth and began walking around the Lutheran church. He wanted his neighbors to support the initiative and began to loudly lobby for their votes. Within minutes, the poll workers were yelling at his mom, reminding her that what her kid was doing was illegal. The ballot initiative lost by just a thousand votes. Ricketts was left with the feeling that if he'd just worked a little harder and convinced a few more of his neighbors, the measure might have passed.

When his mother fell ill in 2010, Ricketts was devastated. For the next several months, he flew back home every few weeks, splitting his time between blocking attacks from Republicans in DC and sitting at his mother's bedside in Seattle. When poetry month came in April, he set up a blog with his sisters to share a poem a day with their mom. She did not get to read it for very long. She died before the end of the month.

In the decade that followed, Ricketts thought back on that painful era with regret. Unknowingly, he'd been away from home in his mother's final years. He had justified the absence because the work was so important. He was supposed to be helping to pass the biggest climate law in history. But when that never happened, the failure hung over him. He couldn't travel back in time and make a different choice. But if he succeeded the next time, maybe the story's ending could be rewritten. He needed to finish the job.

By the time we met in late 2019, Ricketts had grown up. He now had a two-year-old son, a decade working in politics, and expertise in climate policy. After wrapping up the Inslee campaign, he decided to use his excess energy to take on two jobs rather than one. Ricketts joined the Center for American Progress to help build new climate coalitions.

Plus, he set up a nonprofit called Evergreen to keep working with staffers from the Inslee campaign. Rather than trying to elect a politician, they would work to pass a climate law. When he asked me to join that effort, I was thrilled. This was exactly the kind of collaboration I was looking for, giving me a foothold in the policy world. At Evergreen, they'd been trying to coin a term for a new paradigm that could rival carbon pricing. "Standards, investments, and justice" was the best they'd come up with. It wasn't exactly a catchy phrase, but Ricketts wasn't a communications pro—he was a policy nerd.

Like this slogan, his memos were legendarily long. His motto was "Why use ten words when one hundred will do?" The approach had driven colleagues and editors to madness. But it was also effective. He was good at getting ideas on paper. By the time Inslee dropped out of the race, Ricketts and others on the campaign had written 218 pages of climate policy.

The more I read the Inslee plans, the more compelling I found them. By living through Obama's Recovery Act and then the Waxman-Markey failure, Ricketts had built up an encyclopedic knowledge of climate policy. A decade later, he was going back to these ideas. The focus in climate policy had always been on mid-century goals. But Ricketts was thinking about what could be accomplished sooner, by 2035 at the latest. It was a smart approach. Starting in 2020, 2035 was the halfway marker to 2050, with 15 years on either side. This cut the timeline up into something measurable in our human lifespan. Both Ricketts and I would be almost 50 in 2035. By 2050, we would be nearing retirement. We could measure our lives by how much progress we made toward these goals.

The South Carolina primary took place on the last day in February 2020. Joe Biden triumphed by a wide margin, and it was suddenly clear that he would clinch the nomination. Ricketts and I knew it was time to get yet another campaign to adopt some of Inslee's ideas. Together, we wrote a memo that argued the Biden campaign should put out a second, more ambitious climate plan. We said it would bolster his credibility with progressive young people—a constituency he was struggling to win over.

We arranged a meeting with some of Biden's campaign staffers to pitch them on this idea. They proposed we meet up in person, in Philadelphia, in early March. But with each passing day, I felt less certain that I should be flying across the country. We all decided to cancel and stay home. The coronavirus had morphed from something buried on page ten of the newspaper to the top story of the day. What we didn't know then was it would stay that way for the next two years. In the coming weeks, I barely went out, like everyone else who wasn't a frontline worker, and watched in horror as the number of deaths climbed.

But over those first few months of the pandemic, a curious thing happened. People weren't driving their cars and they weren't taking planes. Pollution fell. The air in cities got cleaner. At first glance, it looked like an affirmation of the idea that individual behavior changes could make a dent in the problem. And yet, pollution fell by just 7 percent. I knew that number—that landmark 2018 scientific report had said we needed to cut carbon pollution by 7 percent *every year* until 2030 to limit warming to 1.5°C. During the pandemic, we unintentionally slashed our carbon footprints, exactly as the oil- and gas-funded ads told us to, and we had little to show for it. What's worse, that 2020 pollution cut quickly reversed as people went back to their old lives. None of the infrastructure had changed, so the pollution returned. The pandemic only reinforced my belief that behavior change was a distraction. We had to change policy.

As the lockdowns began, the Biden team reached back out and suggested we meet online. It was the first of many important meetings I would participate in from my home in Santa Barbara. They got to the point quickly, grilling us about the feasibility of cleaning up the electricity system within 15 years. Clearly, Biden and his senior campaign staff were considering adopting Inslee's target of 100 percent clean electricity by 2035. After the meeting, Ricketts called. Sensing an opening, we agreed to draft another memo fast. We wrote up our arguments, sent them over, and waited. We were anxious to hear if they would adopt our ideas, but the Biden campaign was taking its time with its second climate plan. After their first iteration received such a poor reception, they wanted to get it right. Although we stayed in touch with the campaign, we wouldn't find out for months whether they had taken our advice.

Despite the outcome in caucus after caucus, Senator Bernie Sanders did not concede the race until April. He wanted to ensure that his progressive ideals did not get lost. Surprisingly, Biden shared Sanders's perspective.

Traditionally, candidates run to the center during the general election. But Biden wanted to change that pattern. He believed that his job was to put the party back together and make sure progressives and young people could see themselves in the Biden campaign.

In an attempt at consensus after a long, fraught primary, Sanders and Biden established a series of unity task forces that May. Representative Ocasio-Cortez led the climate task force, and Varshini Prakash from the Sunrise Movement also participated. After months of work, they released their plan. Right at the top of the second page was Inslee's 2035 clean electricity goal.

Still, months passed without a new climate plan from the campaign. Years later, I would learn that the delay came from Biden himself. He was personally spending a ton of time on his new plan—probably more than was ideal for someone on the campaign trail. But he didn't want a platform with big, delusional numbers that he could never deliver. He wanted to be able to stick to his word.

Dozens of groups were pressuring the campaign to go big on climate. A coalition of climate organizations and grassroots groups sent a memo with many of the same policy asks Ricketts and I were making. The Sunrise Movement was still rallying young people to pressure Biden. And the campaign was meeting directly with environmental justice leaders to get their ideas for what was needed.

But Biden knew that getting any climate law passed would require a bigger coalition. The climate and environmental justice movements alone wouldn't cut it. He'd seen how Obama's climate plans had alienated organized labor, and he didn't want to repeat that mistake. To get a bill through the Senate, he'd need labor on his side. And that would require him to focus on carrots rather than sticks. It was a lot easier to sell a law that created jobs than one centered on regulation. Biden's plan would create jobs. Good-paying jobs. Unionized jobs. The farther into the details Biden got, the more energized he felt. He knew he was still the climate guy.

But I didn't know any of this at the time—it was all happening behind the scenes. I was just waiting to see if they'd ever release their long-promised new plan. Rather than stay glued to my computer for yet another month, in July, I headed up to the Sierra Nevada for a few days of backpacking with my husband. He was also an academic working on climate policy and was equally eager to take a break. We slept the first night in the foothills on the eastern side, then hiked Kearsarge Pass to Charlotte Lake, where we pitched our tents. We took a break from constantly wearing masks, swam in the turquoise lake, and sang off-key versions of our favorite songs a cappella to the blinking stars. Sleeping under a canopy of whitebark and lodgepole pines, my body calmed down for the first time in months.

When we emerged from the mountains a few days later, I turned on my phone and discovered a stream of missed texts. The Biden campaign had released their new climate plan while I was offline, and they were anxious for me to post a review. I laughed at the absurdity—why did it matter so much what I thought? But after a year of trying to influence policy, I was starting to succeed: decision-makers were listening to what I had to say. I looked up the plan, read it quickly, and found they'd adopted the goal of 100 percent clean electricity by 2035. Before we'd even pulled out of the parking lot, I wrote a quick assessment of the new plan—this time glowing—and posted it online. Biden had won the "most improved" award.

Over the coming months, Ricketts and I would begin writing a detailed roadmap for how Congress could meet Biden's new goal. In the process, I met dozens of people who shared our vision, people I would work with in the coming years to try to get it done. When I began calling up experts to ask them questions, no one thought it was possible to move the idea through the Senate. But by the fall of 2020, we'd figured out six ways we could get a clean energy requirement passed within a big investment package.

As the months passed, I learned how activism works. It requires a person to step out of their day-to-day routine and use the democratic system to try to drive change. There's no one way to do it. It's improvisational and spontaneous, like jazz. Someone starts making up a melody. Others jump in, complementing the tune, trying to add to the rhythm to create harmony, to make the sum of all these sounds greater than each part. Sometimes people bump into one another, the notes clashing. Sometimes the song falls apart. Other times, a piece of music emerges that no one expected. Over the next two years, the campaign for a big federal climate law was like that: a blend of people trying to work together to make something bigger and more beautiful than they could on their own. Each person brought their unique talent to the ensemble, creating beautiful music from different sounds.

In those early days, it was easy to dismiss the work climate and environmental justice activists were doing. The conventional wisdom is that campaign promises don't matter. All the words spilled on debate stages, all the fine print in policy platforms: empty, hollow promises that will never be implemented once a candidate is in the executive suite.

But I still hoped we'd prove that theory wrong. Working with these nascent coalitions, Ricketts and I had dug an idea out of page 14 of a policy document for a candidate polling in the single digits and helped get the winning candidate to adopt it. Eventually, we would need to get the president to endorse it and get it into a bill on the Senate floor. First, Joe Biden just had to beat Donald Trump in the election.

4

RECONCILIATION

By the time Adrian joined Senator Schumer's staff in early 2019, the presidential primary was underway. Candidates were dropping climate plans every week, media coverage was rising, and so were protests. Schumer was watching the crescendo of activism and could see these kids were passionate. He'd been a politician long enough to know that in politics, passion mattered.

Adrian began to feel like the lost decade might finally be coming to an end. But the Waxman-Markey bill's failure in the Senate still haunted him. He'd learned two big lessons from that period. First, Republicans couldn't be relied on. There was no point in developing a strategy that hinged on a single Republican vote. And second, climate action had to be the Democrats' number-one priority. If it fell to a third-tier issue, behind economic recovery and healthcare, Congress wouldn't get it done.

There was also the problem of the filibuster—an arcane Senate procedure that allows debate to continue indefinitely unless 60 senators vote to end it. Traditionally, senators actually held the floor for hours to derail legislation. In 1957, one senator talked for 24 hours straight to stall the Civil Rights Act. But by the 1970s, senators could block a bill just by opposing it—they did not actually have to stand there and talk. This meant that without a supermajority of 60 senators, bills went nowhere. And in an evenly divided chamber, there was no way to convince 60 senators to vote for climate action.

Still, with just 50 votes, the Senate could move a package through budget reconciliation. This process could only be used once a year, and the bill could only address budgetary issues. First, the Senate Budget Committee created a resolution that instructed committees to develop legislation, with clear boundaries for how much government revenue could be spent or raised. The bill would have to follow the "Byrd Rule"—reconciliation procedures outlined by Senator Robert Byrd in the 1980s. The thorniest requirement was that spending not be "merely incidental" to the policy change, since reconciliation was only supposed to be used to adjust the budget. What did "merely incidental" mean? That was in many ways the whole ballgame. Whether or not a given bill followed the Byrd Rule was up to the Senate parliamentarian—an expert on arcane parliamentary procedure who acted as an umpire. Drawing on precedent, the parliamentarian decided what policies weren't "budgetary" enough for reconciliation and cut them from the package. Anything else could pass with a simple majority of 50 votes, with the vice president casting the tie-breaking vote. In these cases, debate was limited to just 20 hours, circumventing the filibuster.

Adrian decided to make it his life's mission to ensure the Democrats passed a climate law through reconciliation. It was the only way to do it. Others were coming to the same conclusion. In biology, "convergent evolution" describes how unrelated species evolve similar traits when exposed to similar environmental pressures. Poisonous frogs, for example, have independently developed bright orange colors to warn predators. And while it's common to find a cactus and a euphorbia living happily together in a drought-tolerant garden, their ancestors developed sharp, spiky appendages while living on different continents. Adrian could see the same pattern playing out with climate policy. Staffers in the House and the Senate, people working on presidential campaigns, and climate activists were all coming to the same conclusion: reconciliation was the only path.

The House was already preparing for a big climate bill. After the sit-in with the Sunrise Movement and Ocasio-Cortez, Speaker Pelosi had agreed to set up a Select Committee on the Climate Crisis. Representative Kathy Castor from Florida led the effort, and committee staff were busy compiling ideas for the future bill. Castor was the right person for the job: a former environmental lawyer, she had boarded up her home to escape Hurricane Irma in 2017. There was also an analogous effort in the Senate, led by Senator Brian Schatz from Hawai'i. A long-time climate hawk, Schatz was also the ideal pick: he was well-liked across the caucus. He didn't have a big ego and went out of his way to help colleagues achieve their legislative priorities.

If the Democrats took the Senate majority and Biden won the presidency, Adrian knew there would be a chance for a reconciliation bill. But if that happened, every policy staffer across Congress would try to advance their priority using this same strategy. Adrian needed climate policy to be first in line. He needed to make a climate bill feel inevitable. He needed to build an echo chamber, repeating it to anyone who would listen until they started repeating it too. And Adrian brought his competitive streak to the task. He might be new to the leader's office, but he could put in long hours. If pressed, he could work harder than anyone else.

But his work ethic came with a price. Adrian started to come home later and later. Eventually, his wife was the only one picking up their daughter from daycare. Weekends were the next to go. Schumer had a standing meeting on Sunday mornings, the only time everyone in the office was free, and it could stretch for three hours. Before he took the job in Schumer's office, Adrian liked to garden, exercise, fix things around the house. Now he didn't do any of that. The work was beginning to consume him.

In his obsessive way, Adrian kept looking at the world backward. Attempting to reverse engineer success, he would think about where he wanted to go, and then ask himself how he could get there. He applied this idea to reconciliation. Even though the process had barely begun, Adrian was already fixated on the final hurdles he'd need to overcome to get this hypothetical climate bill passed in the Senate.

He knew the package was unlikely to include a carbon tax—even though economists had pressed Congress for decades to pass one. But, like many others in the Democratic Party, Adrian had soured on the idea. It was deeply unpopular and worked better on paper than in practice. It was time to focus on carrots rather than sticks.

Instead, this reconciliation bill would involve a large number of smaller climate investments, making it more like a symphony than a solo performance. This had its benefits and drawbacks. The bill would be less vulnerable if one policy was cut, and more senators would be able to see their pet project in the package. But a piecemeal approach would make selling the bill much harder. How would anyone understand a package made up of a thousand tiny parts?

Adrian knew they would need something to prove to the world that the final bill was ambitious, cutting pollution at the necessary scale. An emissions model would be key to selling any package. But this wasn't the kind of thing the Senate normally did. No one had ever built a model like that in Congress before.

Adrian found an equally motivated partner in his new colleague, Tim Ryder. Tim was almost a decade younger and looked it, with boyish features and hair that was four inches longer than other men in the office. They had a lot in common. Both had planned to be ecologists until they learned about climate change and realized they could make a bigger contribution working in politics. In Tim's case, he spent four years working with Al Gore on those climate slideshows, eventually building a close relationship with the former vice president. Then Tim found his way to Congress hoping to maximize his impact. Betting that a moderate Democrat would be the pivotal vote on a climate package, he took an entry-level job with Florida Senator Bill Nelson in September 2018. It was a bad bet. Nelson lost reelection two months later, and Tim was out of a job.

But in the new year, an entry-level position opened up in Senator Schumer's office. Even though it wasn't a policy job, Tim applied anyway, hoping for a chance to work for the leader. He was hired and started in May 2019, just a few months after Adrian. Initially, Tim was tasked with responding to letters from the district. Thousands and thousands of letters. He figured that if he could get all his correspondence work done, he could spend his extra time on climate policy. So Tim devoted himself to the task, becoming extremely efficient at managing correspondence. Within four months he'd replied to 80,000 letters, clearing the office's perennial backlog.

Tim was promoted fast, becoming Adrian's deputy by that fall. Together, they started to build the model that Adrian envisioned. It was a secret project that had to stay within the office. Tim wasn't allowed to talk to outside groups about what he was doing, making it harder to get help when he got stuck. With no prior experience, he taught himself to code just by watching YouTube videos. He could see in his head so clearly what he wanted to make, but knew he hadn't done it yet. He'd have to keep working on it for months, or even years.

Adrian and Tim loved working on their shared project, devoting hours every day to the task. As the presidential primary turned into the general election, building their model kept them sane. If the Democrats failed to win the White House and the Senate, it might turn out to be a theoretical exercise. But imagining their work might just prove useful gave them their daily dose of optimism. And with Trump in office, there wasn't exactly a lot of other climate policy to work on.

By summer 2020, a year into the project, they were ready to present the model internally. When they showed their boss Gerry Petrella the latest version, he was impressed. It had become sophisticated, with a complex backend and a slick interface. But it retained a nerdy charm: the website where the model lived read simply "Welcome to the ambition tracker, created by Tim Ryder."

And while Tim dove deep into the data, Adrian and Gerry focused on mastering the intricate parliamentary procedures involved in reconciliation. Before long, they were giving presentations to Schumer's senior leadership on how reconciliation worked. Adrian was fixated on owning the reconciliation process. When he looked around at the other policy teams in the office, he couldn't help but feel proud. "We're fucking crushing this," Adrian told himself. "We're the only ones that have any plans ready to go when we're in the majority. We have a fucking tracker."

He couldn't have known at the time how important their model would become. How it would prove the centerpiece in negotiating the biggest climate bill in American history. How senators would ask for yet another briefing, demand to look under the hood, question every assumption. For now, it was just Adrian and his friend Tim—two kids building a model rocket in the backyard.

I hadn't spoken to Adrian for more than a decade when he reached out on Facebook in August 2020. He'd seen my recent work on federal policy and wanted to reconnect. Since I rarely used the website, I missed his message for a solid couple of weeks, and then hoped I wasn't too late. I knew Adrian held a central role in the climate policy world. Plus, I liked the guy.

After reconnecting, we set a time to talk over the phone. I told him about the report that Ricketts and I were busy writing, trying to map out all the pathways to clean up the electricity sector in a reconciliation bill. But Adrian was not pleased. Sounding like a stern teacher who was angry about a late assignment, he told me we should have finalized our ideas yesterday. The election was months away, and no one knew who would be the president, let alone which party would control the Senate. But Adrian wanted everyone to be ready.

He held himself to the same standard and was preparing to move as fast as possible if the Democrats won the Senate. He was holding a lot of pieces in his mind simultaneously, trying to iron out the strategy, and the policy, and the messaging ahead of time. Maybe they could have a climate package on the floor in January 2021, right after inauguration day. If the leader's office was prepared, the committees were prepared, and the outside advocates were prepared, there would be no excuse for delay.

Most staffers and advocates saw the world differently and were reluctant to do a bunch of pointless work that would never get used. Plus, the pandemic was still raging. When Adrian gave them the same scolding I got, they probably walked away thinking he was nuts.

But Adrian's message hit me like a truck. The next time Ricketts and I talked, we agreed that we had to work twice as fast to finish the report before the election. By October, we'd picked up our pace and were sending specific pieces to experts for review. If—or rather *when*—the Democrats won the Senate that November, we would be ready to release our plan. All the polling showed they would have at least 51 seats, maybe even 53. We were looking forward to the day that Schumer finally became the majority leader.

5
ELECTION DAY

On the morning of September 9, 2020, more than 20 wildfires were burning near San Francisco when Sonia Aggarwal went to the hospital. It was her due date, and the doctors wanted to check on the baby. She emerged hours later, still pregnant, to ominous orange skies. It was noon, but it looked like the sun was setting. The smoke was so thick it cloaked the sky in a dystopian hue.

Sonia knew this was climate change. Scientists had predicted that fires like these would come, more each year, if we continued burning fossil fuels. She'd made it her life's mission to do everything she could to cut pollution. But she wasn't sure if all her hard work would pay off. "Between COVID and the wildfires and Trump—we were trying to hope for a better future," Sonia later reflected. "But at that moment it felt like everything was burning." With less than two months until the election, it was a scary time to bring a newborn into the world. No one knew if Trump would win again, imperiling democracy and making climate action all but impossible for the next four years. There had already been far too many years of inaction.

By then, Sonia was in her mid-thirties and had worked on climate for a dozen years. As vice president at the nonprofit Energy Innovation, she built complicated models to help policymakers understand how to quit fossil fuels. From China to India, her tools were shaping policy across the world. But it was a job she wouldn't hold for long.

She became pregnant in early 2020, right before the pandemic hit. At first, it seemed like the best life hack ever—working from home meant that she could wear whatever she wanted, and no one would suspect she was expecting. But with time, she started to see the downsides. She had to attend every ultrasound alone—the medical staff wouldn't even let her husband listen in over the phone. And some of the scans were scary. At 20 weeks, the doctor thought her growing baby might have a hole in his gut, requiring her newborn to be rushed into surgery immediately after birth. Sonia absorbed this devastating news alone in the examination room. When the appointment was over, she walked out to her car, sat in the driver's seat, and sobbed.

Despite this early scare, the rest of her pregnancy went well, and over time Sonia relaxed. She started to believe that her baby would be born healthy. Whether or not the world he was coming into would be safe was another question.

Throughout her pregnancy, she volunteered for several Democratic presidential campaigns. Once Biden emerged as the front-runner and began developing a second climate platform, she briefed his team regularly. They were impressed by her work, including a report she'd published that looked at how fast the US electricity system could clean up. The campaign wanted to understand her analysis in detail. Sonia encouraged them to adopt a goal of 100 percent clean power by 2035—the exact same thing Ricketts and I were pushing them to do.

Sonia and I met late that summer—remotely, of course. We hit it off immediately. Just minutes into our first call, I said it felt like we'd been living parallel lives. She was just a few years older than me, and we grew up a mere 200 miles apart, both living on the shores of the Great Lakes: Sonia was in Ohio and I was in Ontario. In our teens, we'd spent lots of time in the wilderness, sleeping outside alone. Then, in college, we'd taken courses on Buddhism and read Sylvia Plath. Now, we were both obsessed with the idea of cleaning up the electricity system by 2035. There were differences, of course. She had dark, curly hair; mine was fine and light brown. Her dad was from India; mine was born in the United Kingdom. But it was strange how looking at Sonia on a Zoom screen felt like looking into a mirror.

At the time, she was already eight months pregnant. I asked why she decided to have a kid and listened as she explained my own reasoning back to me. Sonia worried that becoming a mother would narrow her focus—from doing good for as many people as possible to prioritizing one child. There were only so many hours in a day, and if she spent them changing diapers, would that mean there was less time for climate policy? I'd never met anyone who saw this question the same way I did. It was like she was living inside my head. But, in the end, Sonia decided to become a mom despite her fears. Maybe she was wrong about the costs of motherhood—maybe her child would deepen her commitment to her work, giving her a constant reminder of what she was fighting for. The conversation stayed with me. Maybe I was wrong too.

A month later, Sonia's son was born, almost two weeks past his due date. During that first scary scan, the doctor had told her he would be born with a giant head and lots of hair. Instead, his head was average, and he was almost bald. But the most important thing was he was healthy.

By then the record-breaking wildfires in Northern California had smoldered out. Politically, it was still a stressful time: Supreme Court Justice Ruth Bader Ginsburg had died, and Republicans were working overtime to ram through their nominee, just in case they lost the November election. In the last stretch of the campaign, Biden focused his message on four big challenges: the pandemic, the economy, racial inequality, and climate change. This gave our issue top billing. Sonia was watching the race from home with her newborn son on her lap. She knew that the result would matter for his future.

Almost a week after the election, it was still unclear which party would lead the Senate, but Joe Biden was finally declared the winner, and his campaign team started preparing to govern. Within days, Sonia received a phone call asking if she would serve on the transition team for the Department of Energy. Even with a six-week-old baby, she said yes.

On the West Coast, the calls started every day at five or six in the morning. The team reviewed what had been happening at the Department of Energy under Trump, trying to get a handle on just how much had been dismantled. From 2017 to 2020, Trump picked away at the department, cutting clean energy programs and boosting fossil fuels. But the Biden transition team didn't want to just rebuild the old Department of Energy—they wanted to reimagine it with climate action as its central goal.

The meetings were long, with few breaks. To nurse her son, Sonia would angle the screen upward. Afterward, he would nap on her chest, the sound of his mother's voice lulling him to sleep. She was trying to do it all: motherhood and policy work. But the hours for both jobs were long. When her work on the transition team finally wrapped up, Sonia felt relieved. She was looking forward to returning to her maternity leave.

Then a representative from the transition team called and asked if Sonia would consider a position in the administration. She could join the Climate Policy Office—the first time such an organization would exist in the White House. It felt like a surreal request. Sonia didn't expect to work in the federal government, let alone the White House. Yet here was her chance.

There were a thousand reasons to say no: it was the middle of a pandemic, she had a newborn at home, and she had just restarted her maternity leave. With the Senate majority undecided, it wasn't even clear that there would be an opportunity to legislate. If she took the job, she might sacrifice time with her newborn, time that she would never get back, and have no climate law to show for it.

And yet Sonia felt a personal responsibility to do everything she could to stop the climate crisis, a motivation that motherhood only deepened. A job in the White House came with good odds to shape history. She knew that she couldn't pass up even the smallest chance to help enact a national climate law. The administration's leadership choices were also inspiring: Gina McCarthy would run the climate policy office, and Jennifer Granholm would head up the Department of Energy. After a grassroots pressure campaign led by my friend Julian NoiseCat, Deb Haaland would lead the Interior Department—the first Native American in the cabinet. These picks made Sonia think the Biden team was serious about climate. She took the job, ending her maternity leave early for the second time. As Sonia would tell me years later, "I really didn't know what I was getting myself into."

On January 5, Raphael Warnock and Jon Ossoff won their elections in a runoff in Georgia. They became the 49th and 50th Democratic senators, giving Schumer the narrowest majority. Sonia was thrilled.

That elation was short-lived. The next day, the nation woke up to an insurrection. Sonia watched in horror as an angry mob of thousands descended on the Capitol. At least 174 police officers were injured, and several died in the wake of the attack. She had never thought a job at the

White House might be dangerous. But as a new mom, watching the violence at the Capitol, she felt scared. She'd seen violence like this before.

* * *

Sonia did not always have the best memory for events from her own life. But she could vividly recall the first time she'd heard about climate change. It was 2002, and she was a freshman at Haverford College, studying physics. One afternoon, in an astronomy class, her professor started talking about changes to the Earth's atmosphere. He explained that carbon dioxide had been building up for over a century. This greenhouse effect was warming the globe, leading to far-reaching impacts that scientists were still trying to grasp. But the point of the lecture was to talk about distant planets' atmospheres. After she left the room, Sonia felt disoriented: what he was saying mattered enormously to *this* planet. She couldn't understand why she hadn't heard about global warming before. It seemed too important to be reserved for one lecture in an astronomy course.

Sonia decided she needed to learn more, but there were no courses at Haverford on the topic. She started searching the internet, looking for textbooks, and even completed an independent class on climate. Quickly, Sonia realized that fossil fuels were the problem, and that we needed to rebuild our energy system from scratch. Few people back then knew exactly how that would work. When she came across a newly published textbook called *Sustainable Energy*, she read it like scripture.

After graduating in 2006, Sonia knew what she wanted to do. She had to work in sustainable energy. The problem was, she didn't really know anyone in the field. Her dad, the dean of the University of Akron's business school, told her that if she wanted a job, she would need a business card. So she designed one, typing "Sonia Aggarwal, Sustainable Energy" and her contact information into a Word document. Text alone seemed too plain, so she added hand-drawn illustrations of a wind turbine, solar panels, and a corncob. She didn't want to spend extra money or use a bunch of colored ink chemicals, so she printed the cards in black-and-white, cutting them out one by one. For the finishing touch, she grabbed her art supplies, adding some gray to the turbine, blue to the panels, and yellow to the corn. Shockingly, those business cards did not land her a job in the industry.

But Sonia's hustle paid off in a different way. Searching online, she found a public relations company based in the Bay Area that billed itself as the communications firm for Silicon Valley clean tech. She decided to email the CEO, touting her science background and passion for sustainable energy. Somehow, she landed an interview. After a thirty-minute call, the CEO offered her a low-paid entry-level job that came with a one-way ticket to San Francisco. Sonia packed her life into two suitcases, and headed west.

Within two years, she'd been promoted to an account executive. But Sonia didn't exactly love the job. Writing marketing materials for solar companies was becoming repetitive. She was more interested in understanding the technology. When one of her colleagues quit the firm in 2008, he called Sonia and told her she should meet his new boss, a guy named Hal Harvey, who had just started ClimateWorks Foundation. He claimed that Hal had raised a billion dollars for climate action. Sonia was skeptical: the organization didn't even have a website. By then, she'd seen plenty of Silicon Valley people selling big ideas with little to back them up.

Her old colleague insisted that Hal was the real deal, telling Sonia to get in on the ground floor of this new organization. They were looking for someone to work on communications, and he'd recommended her for the gig. She should at least consider it. Hal proposed they all meet for breakfast to discuss the role. Sonia was sick of working in PR and tried to get out of the meeting. But Hal insisted, saying she'd come so highly recommended he'd be a fool not to try to recruit her. At three in the morning, just a few hours before their scheduled meeting, Sonia agreed to show up. At the very least, she'd get a free meal out of it.

She arrived at the Marin County restaurant on a Sunday morning in July. When she got to the table, she saw Hal had brought along his adolescent twin daughters and his older teenage son. She was surprised they'd all chosen to join their dad for a weekend-morning business meeting—an unusual choice for any teenager, let alone one dressed like a rebellious skateboarder. But she could see that Hal's kids adored him. They acted like he was the coolest guy in the world. Their assessment gave Sonia pause. If his kids liked him, maybe Hal was a good person who she would enjoy working with. Maybe she should consider the job. She left the breakfast meeting open to the opportunity.

Except Sonia refused to spend the rest of her life working in communications. She wouldn't do it. Couldn't do it. So she wrote Hal a note, saying, approximately, "Thanks, but no thanks and good luck with the project," figuring that would be the end of it. Except he wrote back. If she wasn't interested in a comms job, Hal wanted to know what she *was* interested in.

Sonia was astonished. She'd competed hard for every job she'd had, often losing out in the process. Like the time the vegan gelato shop down the block wouldn't hire her, even though she was overqualified, and vegan. Other times, when she did land the job, it was often a shitty one. As a teenager in the outskirts of Cleveland, she'd spent years repairing cash registers. Later, during her first summer in college, she drove 90 minutes each way to a nuclear plant, where she taught safety procedures to people who had worked at the plant longer than she'd been alive. Sonia was used to taking the job she could get. Plus, it was 2008. The world was sliding toward a recession. With each passing month, work became harder to find. Yet here she was, being asked to write her own job description.

Sonia thought back to the work she'd enjoyed the most and realized it always involved research. In college, she'd spent a summer at the MIT laboratory for ultracold atoms—the coldest known place in the universe. After that, she lived at a remote observatory in Chile, high in the mountains where the air was thin and the conditions were perfect for watching the stars. Sonia loved these jobs: loved looking at the data, loved crunching the numbers. So she proposed a role working on research, and Hal agreed. She would report to his deputy.

Sonia was just two weeks into this role when her new boss said she should go to India. On a one-way ticket. For at least two months. ClimateWorks had commissioned the consulting company McKinsey to help visualize climate solutions. McKinsey had a cutting-edge modeling team in Delhi and world-class researchers in Mumbai. Everyone in the climate world had seen their work: diagrams that depicted a series of ascending rectangles, whose size showed how much pollution could be cut with a given technology, and at what price. The McKinsey graphs were iconic. Sonia's boss wanted her to go learn what they were doing.

Sonia didn't want to go to India for months on end. She had built a life in the Bay Area and had just met someone, the man who would become her husband. But her boss was insistent. They needed their own in-house model to understand what technologies would cut pollution at what price. And since Sonia now worked in research, it was her job to figure out how to make that model. So Sonia got on the flight.

They booked her a room at the fancy Oberoi Trident hotel in Mumbai, right across from McKinsey's office. It was the most luxurious place Sonia had ever stayed, and it made her uncomfortable. She constantly felt like she was wasting money. For weeks, she tried to find an apartment to rent, but her boss told her that wasn't a good use of her time. The location was convenient—after breakfast she could just walk across the street and get to work. They needed her to work fast. Finally, she relented and just stayed in the hotel.

McKinsey's Mumbai team was trying to understand where all the greenhouse gas pollution was hidden across the planet. They talked incessantly about "tons": the basic unit for calculating pollution. But wrapping her mind around the concept was not easy. Mathematically, Sonia knew that one metric ton of carbon dioxide weighed 2,200 pounds. But as an invisible gas, it was tricky to visualize what that meant. Sonia found that knowing some basic chemistry helped. Every time fossil fuels are burned, the carbon inside them bonds with two oxygen molecules as it's transformed into carbon dioxide. Since that molecule is now three times as big, it weighs just about three times more. If a full tank of gasoline is 72 pounds, then the carbon dioxide released when it's burned increases by a factor of three, weighing 225 pounds. This meant that burning one tank of gasoline created one-tenth of a metric ton of carbon pollution. Ten tanks of gasoline led to one ton of pollution.

Like everyone else on the team, Sonia was tasked with hunting for tons. It was her job to figure out how many tons of pollution came from a part of the economy, the technologies that could eliminate it, and their cost. She was assigned the building sector and worked long days in the office scouring PDFs and calling up the few experts that existed at the time, recording her findings in spreadsheets. She didn't know it then, but Sonia would spend the next dozen years trying to answer these same questions.

As the information she gathered alongside the rest of the Mumbai team piled up, it was entered into the model. McKinsey had built it in Excel, a decision Sonia quickly understood was problematic. Anytime she wanted to change an assumption or a variable, she would have to wait six hours for the model to spit out a new answer. To save time, she tried running it overnight, figuring she could just sleep while it ground away. But the model was buggy, and half the time it crashed before finishing, forcing her to start all over again in the morning.

Just over a month into her stay, Sonia came across a serious problem with the model's code. She couldn't figure out how to fix it, and neither could anyone else in the Mumbai office, which was full of researchers counting up tons. It was the Delhi team that specialized in maintaining the model, so she decided to fly up there, just for a few nights, to try to debug the code. On Monday, November 24, 2008, Sonia checked out of her hotel room, left most of her luggage at the reception desk, and headed to the airport. She didn't tell anyone at ClimateWorks that she was going to Delhi. It was just for one or two nights, after all. She would be back by Wednesday or Thursday at the latest. But fixing the model proved more challenging than Sonia had expected, and she decided to stay in Delhi for a few days longer, heading back to Mumbai at the end of the week.

That Wednesday night, two terrorists entered the Oberoi Trident hotel lobby. They set off grenades, shot at guests and employees, and took 143 people hostage. Explosions rocked the building for hours. Reports began to emerge that Americans were being singled out and killed. It wasn't long before Sonia's colleagues and parents heard what had happened—the attack was on the front page of newspapers around the world. Her parents were terrified. They thought, correctly, that she'd been staying at that very hotel for the past two months. And they had no idea that she'd flown to Delhi, missing the attack by 48 hours. When she finally reached them, they were traumatized. As was her extended family in India.

It was weeks before she could go back into the hotel to reclaim her luggage. When she entered the dark lobby, it smelled of smoke. Chunks of plaster crumbled out of bullet holes in the walls. The shattered windows were boarded up. Sonia knew that several of the workers she'd gotten to know were among the 32 people killed. When she approached the woman working behind the desk, she didn't know what to say. What was there to say in the wake of so much senseless violence?

Sonia felt she'd escaped death just because of a last-minute decision. "It made me consider my mortality in a way that most twenty-four-year-olds probably don't," she reflected later. Life came with no guarantees. In an instant, it could be over. The experience only deepened her sense of purpose. Sonia resolved to use the time she had on this Earth to make an impact on the climate crisis.

* * *

On January 6, 2021, these thoughts came rushing back as she watched the violence unfolding at the Capitol. More than 2,000 people pushed their way into the building as they attempted to prevent President Biden from being certified the winner of the election. Someone had even placed bombs at the Democratic and Republican national headquarters. Faced again with political violence and the fragility of life, Sonia resolved to get to work. That afternoon, she sat down and wrote a list of day one priorities—not for the president, but for herself. The White House was a maze, and with just a few wrong turns she could lose her way. "I need to have a crystal-clear focus on the things that I want to get done," Sonia thought. She would only hold this position for so long. She had to make every minute count.

Once she started her job, Sonia kept the list she wrote on January 6 open on her desktop. She tried to return to it every few days. She needed to remind herself that life was short, and she'd taken this job to cut as much carbon pollution as possible. The other reminder was her newborn son.

6
CONCEPTION

On Tuesday, November 3, 2020, I woke up excited. There were two big votes taking place: the federal election and my tenure case. It was an odd choice for a political science department to schedule it that day. Still, my department chair called in the early afternoon to share the news that my colleagues had unanimously supported me. Now I could keep this job for life, if I wanted to. I was thrilled. That feeling lasted around four hours.

Ricketts and I spent the rest of the day working on our paper. We wanted to be ready to hit publish quickly, so we could shape the agenda if the Democrats won the Senate. Polling suggested at least four seats were going to flip, but as the results rolled in, I felt deflated. In Maine, Republican Senator Susan Collins won reelection. And the Democrats similarly underperformed polls in North Carolina and Iowa. In Georgia, it looked like no candidate would get a majority, sending the two races to a runoff. The Democrats would need to win both to get to 50 seats in the Senate, and they were running from behind. When I fell asleep late that night, it wasn't even clear that Biden would win, unless he could hold Pennsylvania.

In the morning, I felt hungover. Ricketts and I were supposed to meet with Senator Markey's team, but they cancelled. Instead, we logged on to our regularly scheduled policy call, theoretically to finalize our report. But everyone showed up exhausted and depressed. Although the Democrats held their majority in the House, and Biden looked certain to win the presidency, the Senate seemed all but impossible.

Ricketts was in the worst shape. He'd spent years waiting for the political stars to align for a climate bill, and they hadn't. With his experience, he could see that without a Democratic trifecta, our strategy was all wrong. "It's time to shelve this paper," he declared. It would simply join the long line of policy plans whose time never came. I argued with Ricketts, insisting there was still value in pushing ahead, but he cut me off. "This whole thing has become a theoretical exercise," he said. What was the point anymore?

That meeting marked the beginning of a dark two-month period. Ricketts stopped taking my calls or responding to texts. Others scrambled to land fancy jobs in the incoming Biden administration. People were jockeying for places in the White House or the Department of Energy, and if you weren't on the transition team, you were no longer worth talking to. I got a few calls asking if I would be interested in a White House position, but since I wasn't yet an American citizen, it was unlikely to work out.

As the weeks went by, I was surprised that Ricketts wasn't offered a job in the White House's new Climate Policy Office or in the headquarters at the Department of Energy. He'd held a prominent role volunteering for the Biden campaign and was beyond qualified. But, like me, Ricketts was hard to control—he liked to be the one calling the shots. And some people in the administration would rather hire a person they thought they could dominate than someone with raw talent.

By the new year, it was clear that Ricketts and I weren't getting new jobs. We would continue our work outside of government, remaining largely uncontrollable, free to perform whatever aerobatics we thought would help. With time, we learned to control each other—much to our mutual annoyance. We became like copilots flying a stunt plane, arguing in the cockpit over what trick we should do next, increasingly exhausted by the forces pushing in on us. But that would all come later. For now, we'd stopped working together at all. That is, until January 5.

On Tuesday night, I watched the Georgia election results live. Within a few hours, it was clear that Democratic candidates Raphael Warnock and Jon Ossoff would both win. And when I called Ricketts that night, he actually picked up. His voice came over the line, ecstatic. We both knew exactly what we had to do: we had to finish the report.

We woke up early the next morning and got to work, finalizing our detailed plan for climate action through budget reconciliation. We barely paused to stare at the violence unfolding at the Capitol that day. Ricketts had worked in Congress for years and knew dozens of people who were there. But as disturbed as we were, we knew we had to stay focused.

Our sprint paid off. Within days, we'd organized our talking points, written an op-ed, and were blitzing the media. By early February, we'd published our report. We made the case that this plan was doable because it was already happening: more than one in three Americans lived in a place committed to 100 percent clean power. We also shared polling that showed more than two-thirds of Americans supported this approach. The details were complicated, but the idea boiled down to one thing: the federal government would need to pay electric utilities for more clean power and penalize them if they kept relying on dirty electricity. We knew that this was unlikely to be popular with the utilities because they hated being told what to do, and they didn't want penalties that required them to listen. But we figured if the carrots were much bigger than the sticks, they might go along with it. And we laid out a number of other approaches if this structure failed, including grants to states that incentivized more clean power and extending the tax credits that had already driven so much progress.

Since the early 1990s, tax credits had brought down the cost of solar and wind enormously. There were two basic types: an *investment* tax credit, which covered part of the cost of installing clean electricity generation, and a *production* tax credit, which subsidized each unit of clean electricity produced. One idea was to move these tax credits to a technology-neutral approach, making them apply to a wider range of pollution-free sources, including technologies that hadn't been invented yet. And while these tax credits were essential, the trouble was they were incentives, not requirements. Ricketts and I also wanted investments that paid for performance, pushing utilities to build clean power fast enough to eliminate their carbon pollution by 2035. We worried that without a requirement, utilities would move too slowly.

We also recognized that our dirty energy system did not affect everyone equally. Disadvantaged communities were more likely to live near a fossil fuel power plant. Others had workers who would lose jobs in the transition. We wanted our plan to prioritize investments into these communities. So we proposed that projects in these locations receive a larger tax credit. We hoped this would spur investment into underserved areas with dirty plants while creating jobs where they were most needed.

Creating good jobs could make the policy more popular and win the support of powerful labor unions. In talking with policy experts, two ideas emerged. The first was to add another bonus for projects that employed union workers and paid them competitive wages. The second was a bonus for domestic content if the materials that went into a project were made in the US. Together, these policies would supercharge clean manufacturing up and down the supply chain. The clean economy would boom, with workers leading the way.

But we also knew that good ideas, no matter how well argued, weren't sufficient. The climate movement had good ideas a decade earlier, but failed to get a climate law passed when groups were in complete disarray. Focusing on clean electricity could open up old conflicts in the movement between those who hated nuclear power and those who argued it was carbon-free. Already, there were fights brewing over whether gas could sneak in as a qualified resource. Disagreement is normal, but if we were busy infighting, we would lose. We needed to organize people who shared our vision if we wanted to get our ideas on the agenda and enacted into law.

On one of our endless phone calls, I told Ricketts I thought we needed to set up a coalition. He said he liked the concept but disagreed with the branding. He'd learned from years of progressive organizing that the word *coalition* implied a level of agreement we were unlikely to achieve, given these schisms over what counted as clean versus renewable energy. Instead, he said, we should build a *group*, where participants had to agree to just two principles: they had to support a clean electricity standard, and they had to believe we could pass it through budget reconciliation. That was enough consensus to make progress.

When we scheduled our first call, we were surprised that dozens of people showed up. We began holding meetings each week, and more kept coming. Together, we debated policy ideas, shared communications strategies, and built a campaign to target Congress. As the months passed, the group grew to over a hundred different organizations.

Clean electricity had momentum—but even that one big idea wouldn't be enough. We also had to electrify everything. Electric cars had a growing market and were well represented with corporate lobbyists. Buildings were a different story. We needed to get homes, offices, and factories to stop using fossil fuels. This meant replacing gas furnaces with super-efficient heat pumps, swapping gas stoves for induction stoves, and getting commercial buildings to electrify their appliances and machines too. But the building sector was far less organized than transportation, and pollution from buildings had stayed stubbornly flat for 30 years. Who, exactly, would advocate for heat pumps?

I started looking around for another group like Evergreen to work with, to create a similar coalition that could push for electrification. After a few meetings that went nowhere, I talked with Adrian. "You should check out this group called Rewiring America," he told me. Six months earlier, they had published a report on electrification that I'd read closely. It was full of witty charts and good ideas.

I took Adrian's advice and set up a call with Rewiring America's CEO, Ari Matusiak. He was just a few months into running the organization but had an impressive résumé that included four years in the Obama White House. Even though we'd just met and I came to the meeting with a rather pushy agenda, Ari was open to hearing my perspective. He agreed that a federal electrification policy coalition was a good idea and offered to staff it with the first folks he hired.

Building these two coalitions, for clean electricity and electrification, was the foundation for all our work that followed. Giant sequoia trees are the largest in the world, towering more than 250 feet in the air. But their roots can travel just six feet underground. Their secret is interconnection. As the years pass, each tree becomes intertwined with its neighbors, forming a dense web of support. They use these connections to share nutrients. And when the winds come, the sequoias lean against each other. In this same way, the networks we built across the climate movement gave us power. They helped us exchange information and weather attacks. They helped us refine our policy proposals, drive the media narrative, and coordinate on congressional outreach. They also kept me extremely busy. My days were packed to the brim.

And then I discovered I was pregnant.

In that dark winter period, when it looked like the Democrats had lost the Senate and I wouldn't land a job in the administration, I decided that if I was going to work from home for yet another pandemic year, maybe it was a good time to try to have a kid. Still, I feared that pregnancy and motherhood would sap my energy, reducing the size of my carbon wave.

This fear only grew when I found myself a few weeks pregnant, just as President Biden took office and the Senate flipped to the Democrats. After a decade of gridlock, we had a narrow opportunity to pass a climate law. More than anything else, I wanted to help ensure that happened. Sure, I was pregnant, but I could keep working. Nothing would change.

Around six weeks in, I put on a mask, climbed the stairs to the second floor, and found my doctor's office. I was going for my first ultrasound appointment, alone. It was a year into the pandemic, and doctors still wouldn't let family members attend appointments. So I sat in the small room, on the exam table's scratchy thin paper, and video-called my husband. He was waiting in the parked car downstairs.

A few minutes later, the technician walked in. A kind, middle-aged woman with children of her own, she explained she would take the first scan of the tiny embryo growing inside me. As she pushed the wet gel onto my flat belly, I saw a grainy black-and-white image appear on the screen beside me. It was like peering at the ocean floor through a sonar machine. I couldn't make any sense of what I was seeing.

"Oh, there's two in there," she said casually.

I laughed. "That's funny," I told her. "That's exactly the kind of thing I would say if I had your job."

She stared back at me, confused. "No, no," she repeated for emphasis. "There's really two in there. We wouldn't joke about something like that."

Twins. My mind went black. I started swearing, repeating forbidden words again and again, like a mantra. Then the tears came. Lying on that cold table, it felt like my life was slipping away from me. I had important work to do, and I didn't know how I could do it with two babies. My husband, watching the scene unfold over his small phone screen, was in shock.

When we got home that afternoon, I had to teach a class. Up until that point, I believed I would be one of the lucky ones without morning sickness. But right in the middle of my online lecture, I felt my gag reflex coming on. I told the students my cat was scratching at the door, went off-screen for a minute, and threw up for the first of many times.

Two weeks later, I returned for another ultrasound. Afterward, the doctor called me into a small side office, where we could sit at eye level and talk. She said she was concerned that the growing embryo might only have one inner amniotic sac, making them mono-mono twins. If that's what I had, the pregnancy would be very high risk. There were lower odds that both babies would be born healthy, since they could accidentally wrap themselves in one another's cords and need to be delivered far too early. Or if one of the babies died in utero, it could cause a stroke in the surviving twin. To avoid these outcomes, if there was just one inner sac, I would need to be hospitalized at 24 weeks for continuous monitoring. She called it "baby jail."

I was referred to a specialist for a higher-resolution scan, to gather more information. But the next available appointment wasn't for ten days. My husband and I spent that time agonizing over whether I should carry the pregnancy to term if the twins were mono-mono, given the higher risk. While I waited, I didn't tell anyone outside of my family I was pregnant because I was afraid of what might happen. I knew I could lose these babies.

This experience deepened my commitment to the right to choose. I started thinking that anyone against women's access to medical care should watch someone they love go through a high-risk twin pregnancy. Seeing them wrestle with life-or-death decisions would make it clear that denying modern medical care to people who need it is absurd.

When I finally got the higher-resolution scan, it revealed the twins were mono-di, with one outer sac, but two inner ones. My husband and I, relieved, decided to continue the pregnancy, hopeful that I would make it close to full-term. It was still quite risky, but we felt like it was a level of risk we could manage.

To cope with the stress and uncertainty, I filled my time with work, sticking to my packed schedule. There was plenty to do: policy reports to finalize, briefings to make, activists to organize. If I didn't stop moving, I wouldn't have to think so much about the fear I felt. I wouldn't have the space to worry about losing the policy or the pregnancy—or both.

As the months passed, I found these two parts of my life became countercyclical. When the pregnancy was going well, the policy's prospects were poor. And when the pregnancy took a turn for the worse, the odds of passing the policy somehow improved. It was a disorienting and strange experience, akin to what physicists call "destructive interference"—when two oscillations move in opposite directions and cancel each other out. The net result was an inability to dwell in joy or grief for any period.

7

MOMENTUM

During the final stretch of Obama's first presidential campaign, Brian Deese woke most mornings, checked his BlackBerry phone, and learned that yet another company was in crisis. It was late 2008, and the financial system was in freefall. In a kind of gallows humor, a senior campaign advisor proposed a drinking game for Brian and his colleagues at the campaign headquarters in Chicago: when the markets closed at 4:00 p.m., if they had fallen by more than 5 percent, everyone needed to take a shot. And then a week came in mid-September when it happened every day. They quickly had to abandon the tradition.

On November 4, 2008, Obama won the election. Just three days later, General Motors announced it was headed toward bankruptcy unless the federal government stepped in to help. Brian was asked to travel back to DC as the crisis deepened. There was just one problem: he had a dog. His mutt, Sampson, was the most wonderful dog in the world. Sure, he was bad with other dogs and made life more complicated, but he loved Brian, and Brian loved him. The thing was, Sampson couldn't fly. For days on end, Brian would wake up in the morning, tell himself he needed to drive to DC that day, only to find that by 10:00 p.m., he was still in Chicago. His days were packed. Eventually, he realized the only solution was to travel overnight with Sampson.

So one evening the pair set off for DC. Brian drove as long as he could before pulling off the highway into a big parking lot. He figured they could just stop there and sleep for a few hours. When the sun came up, he looked around and realized he was at the historic General Motors plant in Lordstown, Ohio. The company had just celebrated its hundredth birthday, but it wasn't clear if it would live to see another year.

As the transition team began sorting through the rubble of the economy, there were too many things to fix and not enough people to fix them. On the list of problems, the collapse of the auto industry was the fifth most important thing, but there were only four senior staff in the room. They assigned the new kid to figure out what to do about GM and the other car companies.

For the next several weeks, Brian felt like the only one awake at the wheel, trying to save this giant company. As the Obama administration spun up, more senior people joined the effort. But Brian continued to lead, standing out for his attention to detail and work ethic. With others on the White House transition team, he put together a proposal to provide government support to rescue GM and Chrysler. In exchange, the companies would have to support auto workers, invest in greater energy efficiency, and undertake much-needed restructuring. The plan worked. Brian and his team helped save the Lordstown plant, keeping several thousand jobs in the community. Across the country, he helped bring back 115,000 jobs to the auto industry.

Those months Brian spent looking at corporate balance sheets and strategy documents taught him a lesson. These car companies had made bad decisions. They'd grown stagnant, stopped innovating, and fallen behind. For a century, car companies had been fossil fuel companies. But the popular Toyota Prius hybrid, and even the niche Tesla Roadster EV, suggested that this was going to change. And Brian could see that America's big car companies weren't ready.

Brian served all eight years in the Obama administration. When he started the job, he was just 30 years old and looked it, even in a sharp suit and tie, and even with a beard that aimed to cover his boyish features. As the years passed, he became an economic expert and deepened his knowledge of climate policy. It was a topic he took to naturally. His mother was an environmental engineer, and she would point out wastewater treatment plants and industrial cooling facilities to Brian when they drove past.

Working in the Obama White House, he started to notice a pattern: climate policy was unintentionally sidelined, again and again. Without a senior person close to the president working on the file, it slipped down the priority list. If Brian had the chance to serve again, he knew he wanted to fix this problem. He envisioned a White House Climate Policy Office, with a strong leader at the helm and a team behind them. This office could sit beside the other policy councils and ensure the problem stayed top of mind for the next president.

But it didn't seem like he'd get that second chance. When Hillary Clinton lost to Donald Trump in 2016, Brian had to figure out what was next. Nine months later, he joined the investment firm BlackRock. An outside observer might think this was a strange choice. With $6 trillion in assets and around a third of it tied to coal companies, BlackRock was the largest coal investor in the world. It was hardly a place for a climate champion like Brian to wind up. But he knew that building clean infrastructure required massive amounts of capital. And BlackRock was the largest money manager in the world. If you wanted to move large sums of money into clean tech and you couldn't work in the federal government anymore, this seemed like the next best place. He became the global head of sustainable investing.

Three years later, after Biden won the election, Brian's name was floated for a big job, running the White House National Economic Council. But some climate groups pushed back. The Sunrise Movement, which Varshini headed up, protested the choice outside the BlackRock headquarters. Articles in progressive outlets painted Brian as a corporate shill who favored an "all of the above" energy strategy.

Brian was surprised by this reaction. He understood critiques of BlackRock as an institution but felt like people were missing the forest for the trees. The fact was, he wanted the White House job to put climate at the center of economic policy. And he had proven his bona fides. His son was born just three days before he got on the plane to the Paris climate negotiations with President Obama. And he was glad he got on that plane because he'd been key to landing the deal. Brian's ability to disarm people allowed him to connect with negotiators from around the world, and he'd proven central to bringing China, India, and Brazil together to reach a deal. After that, he'd worked hard to finalize the Kigali agreement to reduce global production of hydrofluorocarbons. Slashing these super pollutants would cut a whopping half degree centigrade of future warming. When his son wanted to avoid going to sleep, he would tell his dad that if he was allowed to stay up late, it would be "good for the environment," knowing that might prove persuasive with Brian. And if that wasn't enough, Bill McKibben, arguably the patriarch of American climate activism, had officiated his wedding—although that was long before Brian was a big important guy in DC.

Despite the bad press and the protests, Brian got the job. On a cold day in mid-January 2021, a week before Biden's inauguration, he found himself once again driving to DC to help the incoming president deal with an economic crisis. But his life had changed a lot since that first trip. This time he made the seven-hour drive from Massachusetts without Sampson, who had died that fall after a long and happy life. Now he was married with two young kids, making balancing a White House job that much harder.

As he drove, he made work calls and listened to Biden's speech announcing his plan to rescue an economy still in the grip of the pandemic. But the hardest thing about the drive wasn't the multitasking: it was finding a charging station for his new Chevy Bolt. He'd bought GM's electric car a decade after his work to save the company. It was a great car—and a hopeful sign that American car companies were back in the innovation business again. There was already talk that, shortly after the inauguration, GM would announce their plan to manufacture only EVs starting in 2035. All Brian's hard work had paid off. But if more people were going to drive electric, he knew it couldn't be this hard to find a charger. Once he started in the White House, he would need to deliver on Biden's promise to build 500,000 EV chargers. Brian wasn't the rookie anymore, but he was still the guy who could fix problems.

On January 20, 2021, President Biden took his oath of office. Like hundreds of other political appointees, it was also Sonia Aggarwal's first day on the job. On paper, she was working in the White House Climate Policy Office. In practice, Sonia remained at her desk in her basement home office in Oakland. The pandemic was still killing thousands of people every day, and almost no one was working in person in DC. A remote job had its upsides: it was much easier to nurse her newborn between meetings. Sonia had become friends with her next-door neighbors, who didn't have children but loved kids. They would come over most workdays and help her husband take care of the baby. The arrangement meant she could just walk upstairs and see her kid in the middle of the day.

When she sat down at her computer that first day, Sonia didn't understand the intricate structures nestled inside the White House. She didn't know the purpose of the Domestic Policy Council or the National Security Council, or how they related to her work in the Climate Policy Office. And, in truth, no one did. The climate office had just been invented. Plus her job title was "senior advisor"—hardly a clear description. She noticed that another person on the team held the same job: Jahi Wise. Maybe he knew what they were supposed to be doing.

Jahi joined the Climate Policy Office's kickoff call from the maternity ward of the hospital. He was sitting on a '90s-patterned couch that converted into a bed, his dark beard less trimmed than normal, his wife and days-old baby just offscreen. Jahi's colleagues were surprised he came at all, but he didn't want to miss meeting the team. Gina McCarthy, the national climate advisor and the team's leader, peered in on the scene through the small rectangle on her computer. An unfiltered Bostonian with a thick accent, she told Jahi the secret to success in the chaotic coming years: he was going to need to be very specific about what parts of his life he had to keep under control, and just let the rest slide. If he wanted to be a good father and work this job, kids' toys on the floor or dirty laundry would have to be secondary problems. This advice stayed with him like a lighthouse, guiding him through the months and years to come.

During that first call, Jahi sent Sonia a message, asking for her number. When they reached each other shortly afterward, they bonded over their newborn sons and new jobs. Although they'd both worked on climate for much of their career, they spoke different languages: Sonia talked policy while Jahi knew finance. Sonia had just published a detailed road map to cleaning up the power grid by 2035, and Jahi had spent the last several years trying to build a national green bank. But they were both campaigners, bringing clear goals to this new job and an understanding that these precious years would be short. They needed to get shit done.

How exactly they would do that was unclear. "What do you think we're supposed to do next?" Jahi asked Sonia. She told him she'd been assigned an executive order to work on, but it was not exactly filling up an eight-hour day. "What are you going to do?" Sonia asked in return. Jahi was a calm, methodical guy. "I'm going to wait for further instructions," he told her, knowing that things would unfold in their own time.

This made no sense to Sonia. She was the kind of person who asked for forgiveness, not permission. "I'm just going to start, figure stuff out, and hope it works!" she told him.

Neither would need to wait long: the work appeared within the hour and never stopped. During the president's first few weeks in office, Sonia and Jahi helped launch dozens of new executive orders: rejoining the Paris Agreement, canceling the Keystone XL pipeline, and creating a new requirement for the federal government to drive clean energy into disadvantaged communities.

After this flurry of activity, the pace slowed. As the weeks passed, they watched the White House leadership and Congress shift their focus to passing the American Rescue Plan. Unlike Obama's Recovery Act, this COVID law included almost no clean energy, climate, or environmental justice investments. Sonia started to worry. She knew that Congress had a bad habit of never getting to the end of its to-do list.

By early March, the $2 trillion American Rescue Plan was finally making its way through Congress, and the White House turned its attention back to the broader agenda. But it soon became clear that they were going to have a problem. On a call, Senator Schumer warned Brian and the rest of the White House leaders that his caucus was becoming uneasy about more spending. Another big bill was going to be hard. Adding to the trouble, a bipartisan group of senators started talking about an infrastructure bill. During the first Trump era, almost every week was supposed to be focused on fixing the country's ailing roads and bridges. "Infrastructure week" had become a running gag in DC, even as the Republicans accomplished almost nothing. Now, the same Senate Republicans who had wasted four years were claiming they wanted an infrastructure package. Brian couldn't tell if it was just a stalling tactic.

President Biden had served in the Senate for decades and prided himself on working across the aisle. He wanted to try the bipartisan route. To figure out their strategy, Brian began to sort the administration's priorities into two buckets: everything the Republicans might support in an infrastructure bill, and all the investments only Democrats would support in a reconciliation bill. Would the Republicans support paid leave for new parents, as a kind of human infrastructure? Probably not—best to leave that for reconciliation. How about electric school buses? Yeah, maybe, and that could fit in the infrastructure bucket. What about removing lead pipes? That seemed like a consensus idea—no one enjoyed drinking poisoned water.

The leftover policies that wouldn't fit in the bipartisan bill were a hodgepodge. Alongside the climate investments, there were healthcare policies, tax reforms, education investments, and social programs like childcare and paid leave. Brian began another round of sorting, creating two more piles: one labeled "climate," and the other labeled "care economy." He believed in the entire Biden agenda—but Brian was particularly keen to make progress on climate. He lived with regret that they hadn't managed to accomplish more in the Obama years. But this time around, climate had momentum. As they continued to strategize, Brian found himself in bizarre conversations with colleagues who were concerned that the climate investments would move too quickly, leaving the care policies behind. The idea of climate policy moving too fast was so foreign, it made Brian laugh.

The White House communications staff was tasked with branding these two slices of the Biden agenda and came up with the "American Jobs Plan" and the "American Families Plan." Brian thought it sounded like they were selling a T-Mobile phone package, not a new deal for the American people. But the communications team wasn't going to listen to him.

With the broad framework finalized, alongside the questionable branding, the White House policy councils started nailing down the details. Their assignment was to get the top line number down to $2 trillion. For what felt like weeks, staff from the Climate Policy Office, Domestic Policy Council, and National Economic Council joined calls, pulled up the latest version of the spreadsheet, and debated. As the meetings stretched hours, they reviewed item after item, arguing over whether it should be $10 billion or $12 billion. Whoever cared the most about a given policy would fight valiantly against everyone else. They usually lost. Sonia found it excruciating.

Eventually, these long calls ended. Then President Biden launched the American Jobs Plan in Pittsburgh, Pennsylvania, on the last day in March, 70 days into the administration. With the first 100 days just a month away, everyone knew the clock was ticking. They needed to move faster if they wanted to get this done.

Sonia could now give her full attention to her second big assignment: helping design the administration's commitment under the Paris Agreement. She had to help set the goal for how steeply the United States could cut pollution by 2030. Sonia was ideally suited for this task because her mind was hardwired to imagine the future. She often felt like she was falling forward into her life as it unfolded, and struggled to recall the past because she gave so little attention to it. Plus, Sonia had a more technical background on decarbonization than her colleagues. She was the only one who'd worked in a power plant.

Since her days in Mumbai, Sonia's understanding of the scale of pollution had grown exponentially. She no longer thought in tons but in million metric tons—what she affectionately called MMTs. In 2005, the baseline year for the Paris Agreement, the United States pumped out more than 6,600 MMTs annually. Her goal was to understand how far that number could fall by 2030. She taught her colleagues a useful shorthand, as they all went looking for tons to cut: reducing 70 MMTs cut the country's pollution by just over one percentage point.

If the White House acted like the head of an octopus, the agencies were its tentacles, gathering information from all directions. Sonia called up the Department of Energy, the Environmental Protection Agency, and five other departments and asked: what do you think we can realistically do this decade? But she found that the civil servants on the other end of the phone were dumbfounded by her questions. They had spent the last four years under siege, watching their colleagues quit or retire early, seeing empty seats gather dust as roles were left vacant. The Trump administration had withered the federal government's capacity. And here was Sonia, challenging these people to dream again. Most people struggled to come up with a hopeful vision.

Instead she found herself, in a peculiar way, turning back to the very model she'd spent years building: the energy policy simulator. Its purpose was to help policymakers, and now Sonia was that very policymaker in need of help. So she navigated to the website, started plugging in numbers, and watched the colorful graphs of carbon reductions appear on her screen.

Her boss, Gina McCarthy, instructed her that the president wanted an ambitious plan. He had become fixated on cleaning up the transportation sector, and kept requesting memo after memo on EVs. Sonia did her best to answer his detailed questions, and then started plugging these ideas into the model. If all the car companies adopted the same plan as GM and stopped selling gas-powered cars after 2035, that would cut 230 MMTs this decade. That figure was lower than she hoped, but there was no arguing with the math, brutal as it was. It would take time for older gas-powered cars to hit the junkyard. Once they did, Sonia knew the pollution cuts from the transportation sector would accelerate.

Gina had also been convening solar and wind companies to understand how quickly they could grow. Based on these conversations, Sonia plugged in 80 percent clean power by 2030, which would put the country on track to a clean electricity system by 2035. That would cut another 860 MMTs. And if the Biden administration rapidly built new transmission lines, they could cut even more. Her most optimistic guess was 1,020 MMTs, around 15 percent of the country's baseline pollution. Electricity could deliver more progress this decade than any other sector. Plus getting heat pumps into buildings, cutting pollution from agriculture, and protecting forests could deliver another 450 MMTs every year.

The industrial sector proved the most challenging to estimate. The scenario she'd built required a massive amount of carbon sequestration by 2030 to decarbonize high-temperature heat for steel production. She did not believe the country could build that much carbon capture, given the technology's poor track record to date. Very little of it had been deployed because capturing carbon dioxide was far too expensive. Only a handful of pilot projects were operating, all with underwhelming results. It looked like a pipe dream, not a solution. But Gina told Sonia it was her job to push the scenario as aggressively as she could. That was how they should be setting the target. If the president set a goal they already knew was achievable, that would defeat the purpose. The point was to have the federal government accelerate change. So she plugged some stretch goals into the model for the industrial sector, and it spit out another 900 MMTs. That part was the least believable to Sonia. But if she squinted, she felt like it was possible.

The total added up to just over 3,300 MMTs cut annually by 2030. Once she had this bottom-up, sector-by-sector estimate, she went back to the agencies and presented her ideas. She challenged these civil servants to take the president's goals seriously and think about how the country could get to a clean electricity system by 2035, when her son would be a teenager. Once they had something to react to, the federal staff found it easier to respond. Together, they crystallized on a target: the United States could cut pollution in half this decade. It was ambitious—a little more than scientists said was required to stabilize warming at 1.5°C.

But it was still just an idea, just some words in a memo she'd written for the president. Sonia still needed to convince her bosses that this was doable. She whipped up the fanciest slide deck she could and presented the proposal to Gina, Brian, and the other senior leaders. They agreed it was well thought out. Within days, the president himself had signed off on the plan. Sonia was ebullient.

On Earth Day that April, President Biden announced that the United States pledged to cut carbon pollution by 50 to 52 percent within the decade. The commitment made headlines around the world. It was one of the most ambitious goals globally, and a massive reversal from the dark days of the Trump administration. Sonia knew it would help guarantee a better world for her son to grow up in.

She also knew that without a climate law, this goal would be nearly impossible to meet. But the president was now focused on the bipartisan infrastructure bill. With each passing day, her anxiety grew. Sonia thought they were wasting their time with the Republicans. She wasn't the only one with this worry. For the next 15 months, Brian spent every day with an existential knot in his stomach. He didn't know if they'd made a colossal mistake spending precious time negotiating with the Republicans, only to lose their one opportunity to pass a climate law.

8

THE BUDGET CHAIRMAN

On Earth Day 2021, Senator Chuck Schumer walked from his office through the winding underground tunnels to the Senate chamber. He pulled open the heavy mahogany door and padded across the olive green carpet to his assigned seat, right at the front. Schumer had worked in this room for more than two decades. Now that he was the majority leader, it was his job to address the nation on days like these. He began his speech by applauding the president's bold new climate goal, the one Sonia had worked on, then turned to Congress's role in getting it done. He needed to deliver a legislative package that would get as close as possible to Biden's target and still pass the Senate with 50 votes. A simple thing to remember; a nearly impossible task to complete.

Obstacles lay in his path. After the Senate passed the COVID bill in March, Senator Manchin started complaining. He said they had spent too much money. And then Schumer watched a group of Republican and Democratic senators become fixated on infrastructure, trying to make it Congress's priority. Schumer refused to let the climate agenda get sidelined. When he spoke on the Senate floor that day in April, his message was clear: "Any infrastructure bill we consider here in the Senate must include green infrastructure, create green jobs, and make significant progress towards the reduction of greenhouse gases." Adrian had crafted these words carefully. Like his boss, he was worried that the Republican push for an infrastructure law would prove a distraction.

Schumer had built his passion for climate action slowly. He had a tendency to focus on what was getting attention, and since 2009, climate had fallen off the front page of most newspapers. When the Green New Deal burst onto the scene a decade later, Schumer came across a poll that showed it was the top issue for young people. This fact lodged in his mind. He started bringing it up regularly with his staff in meetings. Representative Alexandria Ocasio-Cortez, who was at the Sunrise sit-in, no doubt sharpened his focus. She was a political superstar in his backyard. The pundits were constantly speculating that she could challenge Schumer in a Democratic primary for his New York Senate seat. Whether or not this was true, Schumer kept an eye on his left flank. He had to ensure he didn't lose young people.

By the time he took over as majority leader in 2021, Schumer had embraced climate change as a top issue. At the end of January, for his first national TV interview since assuming his new role, he appeared on *The Rachel Maddow Show*, a mainstay for progressives. Midway through the interview, Maddow asked him to list his legislative priorities, and when the camera pivoted to him, Schumer replied with just one word: climate. He talked about his two-year-old grandson and the world he'd be leaving him. Schumer had an enormous amount of power to shape that child's future. How could he sleep at night if he failed?

To show momentum, he assigned his staff, including Gerry and Adrian, to start working with committees that April on the reconciliation bill. Although Schumer was well known for having his flip phone surgically attached to his ear, the truth was the best work happened in person, face to face. Gerry and Adrian knew they needed to return to the office. They'd both been vaccinated for COVID early, at the beginning of the year, alongside members of Congress. But that did not allay their fears. Their kids were not yet vaccinated. And senators were ideal vectors: they flew across the country, and met with tons of people, before flying back to DC. The Republican members fairly quickly stopped wearing masks on the floor, and the chamber became a hot spot for the illness.

Despite these risks, Gerry and Adrian started working in person with the Senate Finance Committee to write the energy tax proposals that would form the backbone of the climate bill. The Finance Committee was like a miniature model of the Senate: it was evenly split with 14 Democrats and 14 Republicans. To get anything passed in committee, the chair, Senator Ron Wyden from Oregon, would need to get every Democratic vote. Which meant that every member could make demands. Senator Bob Casey from Pennsylvania wanted to require domestic content for the clean energy tax credits. And Senator Debbie Stabenow from Michigan wanted consumer rebates for the EVs being manufactured in her state. And Senator Sheldon Whitehouse from Rhode Island wanted a carbon price, or at least an import tariff on goods from countries without climate policies. Gerry and Adrian worked with the committee staff to try to meet these demands—they knew they couldn't do everything.

Meanwhile, Senator Whitehouse expressed some skepticism of the emission reduction estimates that Schumer was referencing, which came from the model Tim Ryder had built. Whitehouse cared deeply about climate action and had devoted decades to trying to make progress on the problem. For the past nine years, he had taken to the Senate floor 279 times, giving speech after speech about the urgency of addressing the climate crisis. With Biden in office, Whitehouse thought he'd made his final speech, and that the Senate would finally take action. But he was a meticulous senator, and he wanted to do his due diligence and have Tim explain how the emissions model worked to a group of MIT scientists. So, Tim went to the meeting, and powered up the ambition tracker—giving them the most in-depth look to date, outside of the Schumer office and the White House. As they looked through the spreadsheets, graphs and results, Tim felt like they were searching for errors that would allow them to dismiss the work as half-assed. After an hours-long session, he left unclear of what they all thought. But in the weeks that followed, Whitehouse expressed less skepticism—it seemed like Tim had convinced him that the model was well done. Maybe Schumer's team was serious about climate policy.

After a month of negotiations, the Finance Committee approved an energy policy package on May 26, on a party-line vote. Now, Gerry and Adrian had to get a resolution passed out of the Budget Committee that would send instructions to participating Senate committees to get started on the rest of the climate package. These instructions would include a top-line spending number for each committee. The parliamentarian—whose job was to make sure the bill followed Senate rules—had made it clear that the Democrats only had one more reconciliation bill they could use that year or the next. That meant they had to get everything they could into one bill. To accomplish all of President Biden's priorities, it would need to engage most of the Senate, touching on healthcare, childcare, taxes, energy, and the environment. This gave the Budget Committee chairman, Senator Bernie Sanders, an enormous amount of leverage over the process.

Senator Sanders was fixated on ensuring that any funding they passed could be spent. While he had a reputation for throwing around big numbers, Bernie only wanted to fund things that were real. With billions for electric buses being discussed, he told his staff to call up American manufacturers and ask how many buses they could actually produce the next year. Then he got on the phone himself and grilled CEOs with his trademark frankness. He did the same thing with proposed spending for electrification, weatherization, EVs, and solar. He thought the federal government shouldn't be writing billion-dollar checks if they were unsure the money could be spent quickly. People needed to see the benefits before the midterms and the next presidential election. That was the only way they would believe in government and give the Democrats a majority big enough to deliver on Sanders's full vision.

A decade earlier, he'd installed solar on his home in Vermont—one of the only senators to do it. He'd even ensured that the panels were made in America. Bernie loved his solar, not just for the savings but for the feeling that he was democratizing energy and tackling climate change with his own little power plant on his roof. He wanted all Americans to experience these same benefits. In 2011, he began cosponsoring bipartisan legislation that would cut permitting red tape and speed up solar adoption. This was the kind of simple policy that he preferred. If citizens could point to ways that the government was making their life better, that improved democracy.

The problem was, reconciliation wouldn't allow him to get this permitting reform passed. The parliamentarian was very strict about only including programs that affected spending. So Sanders's staff tried to come up with ideas for spending money in a way that would help working-class Americans adopt solar. They designed grants that would fund local programs, proposing $10 billion for the effort. The plan was to run this funding through the Environmental Protection Agency, even though the agency didn't have experience implementing programs like these. If it went through the Department of Energy, Senator Manchin, as chair of the Energy and Natural Resources Committee, might screw up the plan.

Initially, Gerry and Adrian found it easy to work with the Sanders team because Schumer was open to a big top-line spending number. He had long fashioned himself as a Roosevelt-style Democrat. If they worked together, they could figure out a framework that might work for both Sanders and Manchin.

Meanwhile, President Biden had started talking with other senators about a bipartisan infrastructure deal, including Republican Shelley Moore Capito from West Virginia. Memorial Day weekend was floated as a potential deadline for that effort, but it came and went with no official progress. By the beginning of June, the bipartisan deal didn't seem to be going anywhere.

On reconciliation, Sanders's team ended up running ahead, releasing a budget framework without Schumer's input. On June 17, Sanders confirmed to the press that he'd put together a plan for $6 trillion. Gerry was annoyed when he heard this. This wasn't starting the negotiation off on the right foot. "We already know we have to get Manchin's sign-off," Gerry ranted to Adrian. "And $6 trillion is not in Manchin's nomenclature."

The next step was getting approval from the rest of the Budget Committee, including moderate Democrat Senator Mark Warner from Virginia. As soon as Sanders's top line hit the press, Warner criticized it. "There are going to be some hard conversations inside the caucus about what's aspirational versus what's real," Warner told a journalist. For his part, Manchin was kinder, saying he would give the progressives the benefit of the doubt and listen before criticizing. And since he wasn't on the Budget Committee, his views at that moment were secondary. But Manchin would eventually be the key. Gerry and Adrian both knew this.

And then Adrian found out his wife was pregnant. He was thrilled his daughter would have a sibling. But when he realized the baby was due in February, he started to worry the bill might not be done by then. When he said he wished the kid could come a little later, his wife was baffled. What did it matter if their son was born before the climate bill passed the Senate?

9

EXIT, VOICE, AND LOYALTY

By March 2021, I was living most days on East Coast time. The pandemic had flattened participation in policymaking. Even though I lived on the West Coast, had never met any major politicians, and wasn't yet a US citizen, I was helping to craft federal policy—all while bedridden. Being on Pacific time gave me an advantage. By two in the afternoon, folks in DC would be wrapping up work, and I could take a nap. No one suspected I was pregnant because they could only see me from the neck up, and if I wasn't feeling well, the camera stayed off. On the weekends, I would recover and stay in bed for two days straight.

Morning sickness was constant. Small temperature changes, like the sun coming out from behind a cloud, would make me gag. Simple acts like showering, brushing my teeth, or even walking to the bathroom were triggers. I stopped moving as much as possible. For weeks on end, I spent most of the day horizontal. When friends reached out, I didn't return their texts. It felt like too much effort.

My doctor explained there were four levels of anti-nausea medication, and it wasn't long before I was on the highest one, Zofran—the same drug chemotherapy patients use. On the maximum dose, I could get out of bed and get through the day. If I stuck a ginger candy in my mouth, wore an anti-nausea bracelet, and walked slowly, I could make it from my car to the doctor's office without gagging.

Six weeks into this misery, I finally broke down and told Ricketts I was pregnant. That day, I was supposed to be leading a 9:00 a.m. policy call, but I was too busy emptying out the contents of my stomach, medication included. At 9:02, Ricketts called. I texted that I would be on soon, and by 9:04, I was running the thirty-person meeting, camera off, tears streaming down my face. After it ended, I told Ricketts I was pregnant. "Well, you did sound kind of off at the beginning of the meeting," he said. For the past two weeks, I'd told him I had the stomach flu.

Ricketts shared that his wife was also pregnant, and she was due just one week after me. This felt spooky, since he and I were also born a week apart. I didn't have a brother, but I imagined this was what it would have been like. Ricketts was always trying to tell me what to do.

We both forecasted when our babies would come compared to the climate bill. September 30 marked the end of the government's fiscal year and was the unofficial deadline for reconciliation. Both of our due dates were in early October. If all went to plan, the babies would all be born just after the clean electricity standard passed into law. Ricketts joked that we should name the three infants after the policy, with first initials C, E, and S. When we realized that these were also Schumer's initials, we laughed harder.

Only fools make plans.

It was my first time working on federal policy, and I had a lot to learn. Unlike Ricketts, I hadn't been a Hill staffer. Sure, I'd spent a year working in the Canadian Parliament, but that was another country. Still, my training in public policy and political science gave me some useful tools, including a bird's-eye view on government. I'd spent years thinking precisely about power dynamics, looking at the forces that upheld our dirty energy system, and cataloging strategies to change it. But, ultimately, I had studied policymaking, not lived it. Understanding the detailed steps for a bill to become law was far more complex than the TV show *Schoolhouse Rock* suggested.

The more I learned, the more the Senate seemed like an aristocracy. Given America's global power, it felt like these 100 people controlled the fate of billions. Between the filibuster and the Democrats' narrow majority, one or two obstinate senators could act like kings.

And even if a senator wasn't planning to hold the planet hostage, they could still have major influence. For years, Senator Tina Smith from Minnesota had sponsored legislation to create a clean electricity standard. I met the senator several times over the phone and found her whip-smart and determined. She was one of the biggest climate champions in Congress, pushing for clean electricity her whole career. In 2017, as lieutenant governor, she led an effort to double Minnesota's clean electricity standard. Although it took years, the Land of 10,000 Lakes would eventually pass a law requiring 100 percent clean power by 2040. When she was appointed to the Senate in 2018, she wanted to continue focusing on climate progress. Once our report was published, Ricketts and I worked with her office on a version designed for reconciliation. Together, we ironed out the details and tried to convince Democrats in both chambers that it was doable.

Although President Biden had committed to 100 percent clean power by 2035, getting this passed into law was a lost cause. The utilities didn't believe it was possible, and if they were required to do it, with penalties for noncompliance, they would fight like wounded animals. So we landed on 80 percent clean power by 2030 as our target—a straight shot from the current electricity mix, which hovered around 40 percent clean, to 100 percent in 2035. This left the thorny question of how to decarbonize the last 20 percent as a problem for the future.

In meetings with Senator Smith's staff, Ricketts and I tried to come up with a name for what we were proposing. Calling it a standard wouldn't work. The parliamentarian often ruled against new policy that wasn't clearly centered on spending. *Program* struck a better tone. But "Clean Electricity Program" wasn't specific enough. At its core, our idea was to pay utilities for performance: if they increased their clean electricity share at the necessary pace, they would get funding to defray some costs and keep customer bills low. And if they failed to make progress, they would face penalties. One night, on a long call, Ricketts came up with a name: the "Clean Electricity Performance Program." Since everyone in DC loves an acronym, we quickly started calling it CEPP. I laughed, telling Ricketts it was pronounced the same as the French word for porcini mushrooms: *cèpe*.

By April, we were meeting with House and Senate committees, answering their questions, and developing legislative text. These conversations were intense, reminding me of a dissertation defense, with informed, tough grilling. We tried to show that we'd developed the policy and thought through the details. During this time, we even met with Manchin's staff to discuss it. And in that initial meeting, it was well received. We felt like we had wind in our sails.

But there was also conflict over whether we were taking the right approach. Other groups proposed variations on our ideas, and we tried to dig into areas of disagreement and find consensus. The Environmental Defense Fund, one of the most prominent climate groups, preferred a program centered around pollution reduction rather than clean energy growth. This would put the policy in the Environment and Public Works Committee's jurisdiction. It wasn't a bad idea in theory, but we thought that Manchin wanted anything that looked like a clean energy standard to go through his committee—a fact he later confirmed in writing. In sometimes tense coalition calls, we tried to build common ground, knowing it wouldn't help to have public disagreements. Other times, we tried to give folks ideas for how to channel their energy so they wouldn't accidentally create havoc.

But the biggest conflict among advocates was the definition of clean energy. In Senator Smith's existing legislation, she had set a benchmark of less than 0.4 metric tons of carbon pollution per megawatt hour—a tight requirement that we believed excluded gas without carbon capture. Other proposals set the number twice as high. But some environmental groups worried it wasn't strong enough and began spreading rumors that our goal was to let gas qualify. This just wasn't true. A gas plant emitted at least 10 percent higher than that standard. For the life of me, I couldn't understand why anyone thought we were trying to sneak in a backdoor way to build new gas. With the Sierra Club, I'd recently published a report calling out the 50 dirtiest utilities for their plans to build 220 new gas plants. That was the problem I believed the CEPP would most effectively solve: stopping new gas. If we just extended tax credits, that would drive more clean energy. But without a requirement on utilities' electricity mix, gas could still surge.

All this conflict started to strain my relationship with Ricketts. Stressed that things could fall apart, we both became more controlling. And we were operating in a political environment where knowledge is power. When Ricketts told me he was meeting with the Senate Environment Committee to talk about our report, I asked to join. He was cagey about whether I could come. And it wasn't necessarily because Ricketts didn't want me there. Committee staff were often paranoid about who they were talking to. Other times, I would call Adrian or Sonia, and Ricketts would be irritated that I hadn't looped him in. He had actually introduced Sonia and Adrian to each other back in 2019, and he wanted to be included any time we discussed strategy. We would bicker over small slights like these, like siblings. But like siblings, we still loved each other.

As conflict grew, my mind regularly wandered to a book I'd read in grad school: *Exit, Voice, and Loyalty* by Albert O. Hirschman. First published in 1972, it looked at how people respond when they are frustrated by an organization. They can choose to exit by quitting a process, or they can use their voice to negotiate toward better outcomes. Which path they took depended on how loyal they felt to that organization. If Sanders and Warren were the saints of the progressive movement, they each embodied one of these choices. Sanders called for a revolution to rebuild the fundamentally broken system, while Warren prioritized working within the system's constraints to deliver progress. These were differences in strategy, not goals, but progressives loved to fight over strategy.

Hirschman's ideas kept coming back to me that spring, as I watched how activists navigated conflict over the clean electricity standard. Some people chose to exit, refusing to participate in improving a policy they viewed as fundamentally flawed. They began attacking our work, even trying to pull signatories off a coalition letter for the CEPP. Other groups chose voice, showing up at meetings to debate the best path forward and push the policy in a better direction. I felt this was more productive, but there was nothing I could do to convince groups to change their strategy.

What I found most frustrating was the tendency to criticize people within the climate movement who were working toward the same goals, rather than our common opponents. Too often, critics refused to recognize the constraints the Senate imposed. Or they forgot that remaining with the status quo would be worse than making some progress. Instead, they narrowed the movement, disparaging people who didn't share their tactics. I started to think that a "true" progressive was a rare, elusive species. No one had seen one in a long time. It may just be extinct.

Despite these conflicts, our collaboration across the environmental movement was working. The clean electricity calls kept growing, week after week. Ricketts and I aimed to keep the group open and accessible, even organizing resources for underfunded environmental justice groups to participate. We had all kinds of policy experts, including economists at Resources for the Future, and engineers like academic Jesse Jenkins. As our numbers swelled from 30 to north of 90, I realized we needed smaller working groups to stay productive. We established three more calls: one on policy design, a second on Hill outreach, and a third on communications. I began attending them all, to refine the policy, coordinate intel from Congress, and build a press narrative.

A big opportunity came when I got asked to go on NBC's *Meet the Press* in late April for Earth Day. I watched very little cable news, so I didn't know a lot about Chuck Todd, other than as a host of presidential debates. When the producer called for the pre-interview, she talked about her teenage daughter who was already a climate activist, saying, "The young people, they really get it."

Going on TV is an intense experience. I had to do my hair and makeup and prepare talking points. As someone who rarely does more than shower, this felt like a lot of extra work. But the talking, that was easier.

That morning, Chuck Todd asked me about Biden's new climate target, and I told him it was bold and ambitious. Then he asked the typical cable TV question: do we even have the technologies to deal with the climate crisis? I told him we had everything we needed to clean up our electricity, buildings, and transportation system. There was an electric vehicle sitting in my driveway. These weren't inventions we were waiting for: the technologies were already here. We just needed to get Congress to pass a law to make them affordable for everyday Americans—from heat pumps to solar panels to electric bikes. But the best part came when I explained that the clean electricity standard alone could get us halfway to Biden's climate goals. Todd, closing out the segment, gave the sound bite of the day: "A clean electricity standard. That's a phrase that feels like it could catch fire."

And it wasn't just the electricity work that was going well. By May, we were also making progress in the building electrification coalition I'd set up with Rewiring America. Throughout the spring, CEO Ari Matusiak and I had begun working closely together. Like me, he was a campaigner.

While in law school, he'd built a coalition that successfully pressured House Democrats to pass a law allowing young people to stay on their parents' health insurance until they were 26. And now he was fixated on swapping out all the fossil fuel–powered machines in America. Saul Griffith, an Australian engineer and polymath who cofounded Rewiring America, had been running the numbers and estimated there were one billion machines that needed to be replaced: furnaces and water heaters, cars and motorcycles, lawn mowers and leaf blowers, and of course giant power plants. Ari and Saul were building an organization to replace as many of these machines as possible with clean electric alternatives. It was an extremely ambitious goal.

Ever since Waxman-Markey a decade earlier, Congress had considered creating a program called Hope for Homes to give rebates to homeowners for increasing their energy efficiency. This wasn't a bad idea in theory—who could oppose insulation? But the problem was, the bill also planned to subsidize gas appliances. Ari, Saul, and I thought this was an outdated approach, and we tried to persuade these advocates and their congressional sponsors to incentivize electrification instead. But they resisted. They claimed that switching to electric machines was too expensive, and we needed to keep funding "efficient" gas appliances. We told them this was a fossil fuel subsidy that made zero sense in 2021. In time, we would learn this coalition had ties to the American Gas Association.

At first, the idea of criticizing energy efficiency felt wrong. When I began working on climate in the early 2000s, I typed the term so frequently that I programmed my own keyboard shortcut: two strikes of the letter "e." Yet I started to see it was co-opted by the fossil fuel industry. Take the Alliance to Save Energy: as climate awareness grew in the wake of Al Gore's documentary, they started running carbon footprint campaigns, getting people to buy efficient bulbs and turn off the lights. This channeled the public toward marginal improvements on the dirty energy system. ExxonMobil, the American Petroleum Institute, the American Gas Association, and dozens of other fossil fuel companies funded this effort. Energy efficiency wasn't a threat to them—it was the solution to keep them afloat for decades to come.

Clearly, in 2021, they were still using the idea to prop up fossil fuels. It was one thing to support efficient heat pumps that ran on clean electricity, and quite another to subsidize "efficient" gas furnaces. Unable to change that draft legislation, I worked with the Rewiring America team to design a new program focused on helping middle class and low-income Americans electrify. Senator Martin Heinrich from New Mexico emerged as a passionate champion, even installing a heat pump in his own home. An engineer, Heinrich had worked for years on energy before continuing his career in conservation. Once he was elected to the House in 2008, he worked hard to try to advance climate policy, only to find progress challenging. A decade later, he was a senator excited to go big with a new program. With 121 million homes in America, we knew we needed a lot of money to make a dent in the problem, so we proposed upward of $80 billion for the effort. It was a big ask, but Ari's strategy was to start the negotiation high. Little did we know we would spend the next year fighting with the efficiency advocates over an ever-shrinking pot of money.

And then there were the tax credits—Congress's preferred approach to just about everything. I remember an early electrification coalition call when people started to rattle off the provisions we could harness for electrification: 25C, 45M, 179D. I pretended to have a clue what they were talking about. But who on Earth could remember these obscure combinations of numbers and letters? Clearly, being buried in the tax code had its advantages: policies flew under the radar.

In time I learned that 25C was a residential efficiency tax credit. It did not yet support heat pumps but subsidized gas furnaces up to $150. We wanted to eliminate that funding and instead give a 30 percent credit for heat pumps, including for low-income folks who didn't owe the government enough in taxes. We also proposed that the breaker box upgrades necessary to install heat pumps should qualify too. There was precedent for this: solar panels had been getting support at these levels for years.

But despite our efforts, the conversation moved in the opposite direction, thanks to the efficiency advocates. Their goal was to increase funding for gas appliances, rather than eliminate it. And they weren't interested in funding heat pumps. As our conflict increased, I called up Ricketts, trying to drag him into the fight. He had worked with Heinrich for years and had a personal relationship with the senator. I told him he should be helping. But Ricketts only laughed. He wasn't interested in working on building electrification. Watching us squabble was like watching minnows fighting. He was too busy with the sharks.

By June, CEPP was ready for a public rollout. Alongside Senator Smith, I was scheduled to brief the press and the public at the beginning of the month. But the day before that event, I had to go to the doctor for a twenty-week anatomy scan. The morning leading up to the appointment was booked solid. As I headed out the door with my husband that afternoon, we felt optimistic. The policy work was going well, and so was the pregnancy. We had started to relax and prepare for twins, believing that both children would be born healthy. We even bought a second bassinet.

Unlike many of the previous appointments, my husband was allowed to join me in the room that day. On the screen, the fetuses were just grainy little black-and-white blobs. But they were now big enough to be estimated. Dr. Rask drew lines from head to toe to calculate their length. He measured their heads with circles, their tiny femurs with equally tiny lines, checking each measurement several times. Then the computer fed all that data into a model, which estimated their weight. Dr. Rask declared that Baby B was above average, around the 63rd percentile. And Baby A was tiny: depending on the measurement, somewhere from the 4th to 8th percentile. I couldn't wrap my mind around the discrepancy: they were supposed to be identical. It didn't make any sense.

A talkative, gray-haired man who watched too much Fox News, Dr. Rask was challenging for me to work with. For one thing, he didn't believe in climate change—an opinion he shared often. After finishing most appointments, I'd find myself stuck in yet another fifteen-minute debate with him over whether the planet was warming. Santa Barbara was too small to have multiple clinics specializing in high-risk pregnancy. And Dr. Rask's only colleague was worse—that doctor had a tendency to call her patients old and fat. I stuck with the climate denier and made the best of it. At least he was competent.

Dr. Rask explained that we might be dealing with twin-twin transfusion—a condition where one of the fetuses takes energy from the other. The treatment involved lasering my placenta in half, which was risky. And if it was performed unnecessarily, it would increase our problems.

Although the babies had always been different sizes, it started out small, just a millimeter here or there. I'd chalked it up to measurement error. My husband, on the other hand, noted these discrepancies week after week with increasing alarm. "It's the doctor's job," I told him. "Let them worry about it." But he continued to pore over the medical literature on twin pregnancies. After each appointment, he would stay up late, hunched over his computer, reading yet another academic paper, searching for signs in the low sample sizes. That day, when Dr. Rask's measurements showed a stunning difference between the babies, he was prepared with the right question.

"Do you think this could be sIUGR?" he asked.

"Yes," Dr. Rask replied, as if my husband had asked whether the sky was blue.

I had no idea what they were talking about. When we got back in the car, my husband explained that twin-twin transfusion wasn't the only common problem for identical twins. Selective intrauterine growth restriction, or sIUGR, also came up in 10 percent of cases. Reading more about it that night, I found that medical researchers had different theories about the cause. At UC San Francisco, they called the problem unequal placental sharing. When a placenta formed, an umbilical cord could attach anywhere along its surface. If one of the cords connected in the middle, nutrients would flow easily. If the other cord was connected at the edge, it would be much harder to feed the baby. The result was a large size discrepancy in the two fetuses. And this didn't just happen with twins—even with one child, a marginal cord could result in less nutrients and poor growth.

Still in shock, we spent the night talking about concepts I'd only recently learned: "survival" and "intact survival." There were endless probability charts showing the chances that these children would be born alive and without significant long-term disabilities. It felt like a metaphor for the policy work too. Maybe the bill would survive, but in a mangled, ineffective state. And unlike the pregnancy, there were no peer-reviewed articles to show the odds of success.

I continued to feel stunned the next day during my briefing with Senator Smith on the CEPP. Even though the launch went well, it was hard to focus on the policy given what was going on inside me. Afterward, I called Ricketts to give him an update on the pregnancy. I told him we were thinking of going up to the hospital at UC San Francisco in a few weeks for a second opinion. Ricketts said he was in the same boat—after several miscarriages, his wife was having her own high-risk pregnancy. "We've got about 80 percent odds of ending up with three babies," he told me. "That's pretty good. Better odds than getting a climate bill passed through the Senate."

10

PUMPS

That spring, Sonia drove to DC with her husband, baby, and dog. Their tiny plug-in Chevy Volt was packed to the brim, as was the trailer they were towing behind them. They'd sold their house in Oakland and put most of their stuff in storage, but it was impossible to travel light with a baby. As they made their way across the country, Sonia worked from the passenger seat, her laptop balanced on her thighs, the phone calls endless. Halfway there, they stopped in Ohio to introduce her parents to their nine-month-old baby. They cooed over their new grandchild, who had just started crawling.

In preparation for in-person work, Sonia had been receiving lots of emails. They said that a White House staffer's attire reflected on the president, and professionalism was expected. But with the pandemic and her pregnancy, Sonia hadn't gone into an office for over a year. Her body had changed, and none of her work clothes fit anymore. And besides, her wardrobe was slightly offbeat, with bold patterns and bright colors. Sonia needed new clothes.

With her mother, she went back to the mall she frequented as a teenager, her son falling asleep in the backseat on the short drive across town. She left her mom with the napping baby, softly closed the car door, and headed toward the department store. The bright overhead lights shone on rack after rack of professional clothes. Nothing much appealed to her, but she picked out a couple things and went into the small beige fitting room. As she stared at herself in the mirror, she couldn't help but feel like an awkward teenager again, working her high school job at the Things Remembered headquarters in Chesterland, Ohio. None of these clothes seemed to fit her—they all hung on her body strangely. No sooner had Sonia changed back into her own outfit than her mother called: the baby was screaming. Her mom couldn't figure out how to unclip him from the car seat, and she was frantic. Sonia hurried back, having bought nothing.

By May, they were settled in DC. On the morning of her first day in the office, Sonia was nervous. It felt like the first day of school. She cobbled together an outfit and dug a pair of rarely worn high heels out of her moving boxes. After drinking a coffee, she kissed her baby goodbye, then drove downtown. The parking lot was far away, so she uncomfortably hiked the mile to the security checkpoint, trying to remember how to walk in heels, as if she'd ever learned.

When she finally made it into the building, she saw her colleague Jahi. It was their first time meeting in person. After so much time on Zoom, it felt strange. He was shorter than she'd imagined. Sonia and Jahi got the full tour of the sprawling complex. The White House included the West Wing and the lesser-known Eisenhower Executive Office Building. That was where they would work from shared offices on the fourth floor. The EEOB, as everyone called it, was less iconic than the West Wing, but it was still impressive. Built in 1888, it had thousands of columns, dozens of curved staircases, and elaborate stained-glass ceilings. But Sonia wasn't paying attention to the French architecture. As she walked the halls with their checkerboard black-and-white floors, she glanced at the shoes of each woman she passed. She noticed that a handful were not in heels. It was the permission slip she'd been waiting for: Sonia never wore heels to the office again.

As the weeks went by, working at the White House remained as exciting as that first day. She couldn't believe this was her job. To stretch their legs and talk through ideas, Sonia and Jahi would often take long, circular walks through the EEOB, easily clocking miles without ever leaving the office. But the grand building had its downsides. For one thing, Sonia was always trying to find a place where she could pump. She was still breastfeeding her son, and she couldn't exactly extract milk while sitting across from her colleague in their tiny office. She asked if there was a space she could use, and the Climate Office's conference room was offered. But it didn't have a lock—at any moment, her bosses could just walk in. There was also an official nursing room, but it was as far away as possible while still being in the same building. Sonia worked on the fourth floor in one corner of the 300,000-square-foot complex, and the pumping room was on the first floor in the farthest opposite corner. By the time she walked there, she'd already be late for her next call.

Instead, she shuttled from her office to an empty bathroom stall, shrugging off her clothes and trying not to get her dress wet in the toilet, all while continuing to work on her phone. Other times, when she needed a computer, she would use that conference room, sticking a hastily made sign on the door that read "Please knock." Except there was never any paper or tape around. They didn't have a budget for office supplies, so she kept having to ask other, more established groups to borrow things. Those folks were given funding from Congress, while the Climate Policy Office was not. Even getting the mini fridge she needed to store her milk was a multi-week saga. It wasn't like she could just buy one herself and bring it through security. Nothing was simple when you were working in the White House. In time, Sonia got to know her colleague well enough that she just started pumping in her office. It was easier.

But while she was learning to navigate working motherhood, the job was becoming more frustrating. Every few days, she kept opening her list of ten priorities. The key ones were delivering on the 100 percent clean electricity campaign promise, getting EVs on the road, and building clean manufacturing in America. Sonia knew that none of this could happen without Congress. But the American Jobs Plan had landed with a thud. The only reason she'd taken this job was to pass a climate law—and it had slipped off the agenda.

Instead, the bipartisan infrastructure negotiations were taking up the administration's attention. Senior White House leadership told Sonia to comb through the American Jobs Plan and pull out anything that might be bipartisan. Sonia thought this was a bad idea—if Manchin got everything he wanted out of the first package, why would he agree to the president's broader agenda in a second reconciliation bill? "I never understood the rationale," Sonia told me later. "It was never explained to me."

This delay wasn't just frustrating because Sonia knew the planet was hurtling rapidly toward breakdown. It was also derailing her plans on a personal level. When he was in his thirties, Sonia's husband went back to school to earn an undergraduate degree in mechanical engineering at UC Berkeley. Being a mature student wasn't easy, but he'd graduated, and then landed a job in the solar industry. He'd worked there less than a year when Sonia's White House opportunity came up. He'd offered to quit his new job to take care of their baby and make it all work. Sonia knew this was a big sacrifice, and she felt terrible about it. She told him it was temporary: the climate bill would pass Congress within a year, and then they would leave DC. They could move back to New Mexico, where he was from, and prioritize his career. She'd insisted they bring very little furniture to DC and sign a month-to-month lease because she knew they wouldn't be there for a full year. Her husband listened to Sonia say this again and again. It sounded like she was trying to convince herself more than him.

By the spring, the White House leadership was in an intense debate over how much time they should give the bipartisan infrastructure effort before abandoning it. There were some who wanted to try a solid, fifteen-minute conversation with the Republicans. Everyone could walk away and conclude that Republican Senator Shelley Moore Capito was a nice person, but that nothing was going to come of this. But others in the White House believed that if they failed to pass a bipartisan bill, they would lose support from moderate Democrats on reconciliation, and nothing would pass. These people were willing to spend infinite time on the process.

Brian Deese, now running the White House National Economic Council, found himself somewhere in the middle. He knew that part of Capito's motivation was to delay reconciliation, but she also had a sincere interest in infrastructure. In time, he'd realized his goal was to land as many of Biden's climate investments as possible. If they could get a bipartisan package through with some climate spending, that would increase the total. And that part of the deal wouldn't need to be paid for: as much as the Republicans claimed the mantle of fiscal virtue, they almost never passed balanced bills. They struggled to come up with revenue because they hated raising taxes on the rich and corporations. Brian found this a maddening double standard: Democrats were always being blamed for irresponsible spending, when the truth was Republicans ran structural deficits year after year. But in this case, if it helped his kids grow up on a healthier planet, he'd take that deal any day.

The White House spent the month of May dancing with Senator Capito. For several weeks, she came down from the Hill, often dressed in a bright-colored suit, sat across from the president in the Oval Office, and got right to business. Between these meetings, Brian would head up to the Hill to meet with her staff and talk next steps. He liked her. She was smart, a straight shooter, and did not suffer fools. By then, she'd served in Congress for two decades, and she knew how to make a deal.

But the more they talked, the more Brian realized that she was taking her cues from the Republican minority leader, Senator Mitch McConnell. At the time, the Senate was working to reauthorize the surface transportation bill, which was supposed to happen every five years. And Brian could tell that Capito and McConnell weren't interested in boosting spending above baseline. They just wanted to authorize the same amount as usual and shift how much went into each bucket. Brian knew this wasn't going to work. In meeting after meeting, he kept encouraging Capito to think about things differently. If they wanted to fix the country's ailing infrastructure, they had to increase overall spending. But Capito didn't want to do that.

After the Memorial Day weekend, before the president had to travel to the Group of Seven summit, Brian and the other senior White House advisors decided it was time for a hard pivot. They needed to focus on getting the reconciliation bill done before the August recess. Brian was standing in the Oval Office when the president called Senator Capito to stop their negotiations. Even though it was like a breakup call, there wasn't much emotion or drama. It was a collegial, matter-of-fact exchange where they both recognized they just weren't going to get a deal done and should move on.

But before Brian and the rest of the White House leadership could refocus on the climate bill, Senator Mitt Romney from Utah called. He said they could still get an infrastructure deal done if the president worked with a broader group of Republicans. "Please," he pleaded, "spend a few more weeks with us." After decades in Congress, Biden believed deeply in working across the aisle to get difficult deals done. He told his staff to try it. So Brian and the team got back to work on infrastructure.

Over the coming weeks, a group of five Republican and five Democratic senators met regularly with the president. Gina McCarthy, the National Climate Advisor, lived less than a mile from her office in the West Wing and walked to work most days. When she began to see huge black SUVs parked outside most mornings, she knew the Republicans had come again to talk to the president. It was a hopeful sign that they were making progress. She thought they would get the infrastructure bill done, and then get back to the climate package.

11

NO CLIMATE, NO DEAL

One week after that bad scan, my husband and I headed up to UC San Francisco in early June to understand what was happening and decide whether to continue the pregnancy. It was the closest research hospital that specialized in complex prenatal care. We arrived just after sunrise for a full day of tests. The first round of ultrasounds took two hours, as the technician made 400 measurements of two banana-sized beings. After that, there were several more hours of fetal echocardiograms to examine their tiny hearts, now the size of a quarter.

At the end of the long day, we found ourselves in a conference room, surrounded by whiteboards, waiting to discuss the results with the doctor. A slight woman in her thirties, Dr. Nosci walked in and sat across from us in one of the dozen rolling chairs that lined the large table. She had published research on the condition we were dealing with, and, like the good academics we were, we'd done the reading in advance and came prepared with questions.

Dr. Nosci explained that the problem we were having was physical. When the two embryos formed at the beginning of my pregnancy, one of the umbilical cords made a solid connection at the center of the placenta, while the other was barely connected at the edge. Hence, Baby A was struggling to get adequate nutrition through a marginal cord while Baby B grew normally. There was nothing I could do to fix this problem. Suggestions to rest or reduce my stress were misguided at best and harmful at worst, implying this was somehow within my control, and therefore my fault. But I could not control how an embryo formed any more than I could control when it rained. I had a better shot at controlling the Senate.

Like millions of other couples before us, we were facing a difficult choice late in pregnancy. With twins, it can be risky to protect both babies. The longer they stayed inside me, the better the outcomes. But if one fetus died in utero, the resulting change in pressure could kill or permanently disable the other. Even if both fetuses stayed alive, things could take a turn for the worse at any time. If the smaller baby started to show distress, they would both need to come out, potentially harming the larger child, which was growing well in utero. There were no easy answers.

We grilled Dr. Nosci for an hour straight, asking her question after question about the likely outcomes. I tried to keep up with her medical jargon, taking eight pages of notes. Based on her sample of similar cases, Dr. Nosci told us that 70 percent of the time, people in situations like ours had good outcomes and made it past 30 weeks of pregnancy. The rest of the time, there were mixed outcomes: early births, at the edge of viability, often resulting in lifelong complications. And in a smaller number of cases, one or both babies died. Both my husband and I were professors who taught statistics. We understood probability well. But staring at tables showing week-by-week survival rates was daunting.

It became clear that there were three options. One: do nothing and hope for the best. Two: reduce one of the fetuses. And three: terminate the entire pregnancy. If I chose option one, and continued to just cross my fingers and hope, the doctors called it "expectant management." It was one of the funniest medical terms I'd ever heard. What did it even mean—that the treatment was managing my expectations?

After an hour of questions, Dr. Nosci seemed tired. As a Canadian, I have a tendency to apologize, so I said, "I'm sorry if we're asking a lot of questions." Normally, people reply politely, saying, "No, no, of course not"—even if they're lying. But Dr. Nosci couldn't fake it. She just stared at us. It had been a long day for everyone. We thanked her for her time and left the hospital, grateful for the information she'd shared.

That night, back at our hotel room, we felt reassured. Dr. Nosci knew the data better than almost anyone. And she seemed to think our odds were good. The key would just be detecting any distress in Baby A as soon as possible. Together, we decided to continue the pregnancy, hopeful that both children would be born alive and healthy.

But that optimistic feeling didn't last long, because now there was trouble in DC. Within the hour, I got a call from Ricketts. He had bad news: Gina had given an interview that day. A reporter asked her if climate investments might not make it into an infrastructure deal. Senator Romney had already confirmed that most were likely to be excluded. But Gina went further, specifying that our clean electricity program might not end up in the final version of a climate bill. Ricketts was livid.

I pulled up the article to read it myself and found a terrible headline: "Infrastructure Plan Could Omit Some Climate Proposals." This looked bad. But when I read the full quote, it was clear Gina was speaking off the cuff and didn't intend harm. I could even see her trying to correct mid-sentence, saying she knew our clean electricity plan was necessary and she would "fight like crazy" to make sure it was in the final bill. I spent a lot of time talking to journalists and understood that sometimes you were caught on the record saying something you didn't mean. "I don't think it's that bad," I told Ricketts. Maybe she was trying to suggest it wouldn't be in the infrastructure bill but could still make it into reconciliation. But Ricketts wouldn't hear it. "She shit on our program publicly, and that is not acceptable," he said.

Ricketts went into red-alert mode and began calling, emailing, and texting his extensive contacts. Folks inside the admin assured him it was a mistake, and they would fix it. By the morning, Gina had tweeted a correction, underlining her support for the policy. This was the nice thing about the Biden team: they were responsive to outside pressure.

And the snafu had an unintended effect. The combination of Gina's bad quote and the delayed bipartisan infrastructure package caused a bunch of senators to begin speaking up more forcefully. Senator Smith stayed on message, arguing her clean electricity program was key. Senator Heinrich said he wouldn't vote for a package without climate in it, and Senator Wyden agreed. But the idea really caught fire when Senator Markey told *The New York Times* that he wouldn't support an infrastructure deal without climate at its center, saying: "No climate, no deal."

Across the climate movement, folks seized on this phrase, making it a rallying cry. We repeated it again and again, putting it on websites, banners, and tweets. One week later, Senators Markey and Merkley were standing in front of Congress saying the bipartisan bill wasn't going to happen without climate. By then, Evergreen and Sunrise were running a campaign that asked members of Congress to make the same pledge and stand beside them. Within a few weeks, 14 senators and 14 representatives were on the record saying they supported stalling the infrastructure deal.

It was the Congressional Progressive Caucus in the House that really took the idea and ran with it. The institution had been moribund for years. But when Representative Pramila Jayapal from Washington took it over in 2019, she was determined to use it to wield power. Climate policy was a priority for Jayapal, and she organized her caucus into a wall against the infrastructure package. The progressives in the House worried that passing the bipartisan bill first might put at risk the climate and social investments, including universal preschool, paid leave for new parents, and healthcare funding. If the Republicans weren't going to put progressive priorities into their bill, then the House needed to pass the reconciliation bill and the infrastructure package at the same time.

If anything, Gina's mistake seemed like a net positive. Up until then, Manchin acted like he was the only person who mattered. But with others threatening to veto a watered-down bipartisan approach, the tone of the negotiation changed.

This episode caused me to think more about building outside pressure to support these climate hawks in Congress. In the postmortem of the Waxman-Markey failure, several political scientists had found that the lack of public mobilization for the bill was one cause of its failure. I didn't want us to make the same mistake twice.

The Sierra Club was the environmental group with the biggest membership. When I reached out to their political staff to see if they were planning to ask their members to call Congress in support of the climate bill, the answer was no. They were busy having internal conflict, reckoning with their past, and dealing with leadership changes. Frustrated, I tried to find other groups with big lists, bringing up the idea on our group calls. Many smaller state-level organizations, like the Chesapeake Climate Action Network, were keen to help out. But finding a national partner with a big list was proving impossible. What was the point in maintaining these lists of millions of Americans who wanted to act on climate if you didn't use them to push for the biggest climate bill in American history? When I was complaining about this on a call with a colleague one day, she suggested I reach out to Jamie Henn. Maybe he could help.

Back in 2007, Jamie Henn founded 350.org alongside Bill McKibben and a handful of other activists. He had been responsible for some of the biggest climate actions in history, including the People's Climate March in 2014, which 600,000 people participated in worldwide. He was now running a smaller communications group that aimed to support the climate movement. When we met, we clicked. Like me, he was fast-talking, pragmatic, and willing to laugh at himself.

With other folks from his team, we built a campaign called Call4Climate to help people call their members of Congress. We encouraged activists to make four demands: ensure 40 percent of the total funding went to environmental justice, establish a Civilian Climate Corps, cut fossil fuel subsidies, and create the clean electricity program. We didn't have any funding for this effort, so I filmed myself making a call, being careful to hide my huge belly. Then I downloaded software and edited a short clip before posting it online. Our little campaign would result in more than 15,000 calls to Congress—not bad for no money.

But even with public support, Ricketts and I knew that getting the clean electricity program into a climate bill remained a long shot. By June, we were trying to figure out how much funding was necessary to make it work. There was a tension between failing to get utilities on board with a small program and alienating Manchin with a big program. Ricketts and I wanted a principle for how much money was needed. So we went back to the power-sector modeling Sonia had worked on years earlier at Energy Innovation. And we dug up the analysis that Jesse Jenkins and his colleagues at Princeton were doing. Both suggested that getting to 80 percent clean power by 2030 would require hundreds of billions in new capital spending. After reviewing the data, Ricketts landed on $500 to $600 billion within the decade. We agreed that we didn't want these additional costs to push up electricity prices for everyday Americans, which would make electrification harder. Extending the tax credits might optimistically cover $250 billion. But that left a big gap. To fill it, we debated numbers between $150 and $250 billion for the CEPP. Eventually, we landed on $150 billion as an ideal amount. If that was going to happen, the Budget Committee would need to assign that much spending to the Energy Committee.

Ricketts taught me that when Congress negotiated, the amount of spending always went in one direction: down. Manchin would want to take a pound of flesh from the climate package before he agreed to voting yes—if he ever agreed. We hoped he might take aim at a carbon price. But to be a sacrificial lamb required a policy to first find its way into the package. And so far, there wasn't any kind of carbon price included. Senator Whitehouse was trying to build momentum behind the idea, but it struggled to gain traction. Instead, he turned to critiquing our work, calling us the "CES-only crowd." But Ricketts thought Whitehouse's comments might work to our advantage. If it was clear that a climate hawk like him wasn't interested in our idea, that might actually be appealing to Manchin.

We also heard that Senator Merkley was partnering with Representative Cori Bush from Missouri, a stalwart progressive. They were poised to come out with a letter criticizing the CEPP, saying it needed to exclude gas. That was a worthy point, but the tone of the intervention was negative. Ironically, that might be helpful: if progressives hated our policy, maybe Manchin would like it more.

I began to think that the traditional tools the progressive movement employed, like protests and letter-writing campaigns, were ill-suited to convincing Manchin. When public pressure was applied to him, he didn't seem to respond like other Democrats. He just dug in his heels. And if a policy was seen as a progressive priority, it seemed like Manchin was more likely to kill it.

If we wanted to get climate investments passed this Congress, we needed to think about who he listened to—his top donors and the biggest businesses in his state. So we reached out to corporations, asking if they would speak up for the CEPP. We called utilities all over the country. We talked to big tech companies, clean energy developers, retailers, and more. We found out that many of them were already on it, with a draft letter ready to go out the door in the next few days supporting the policy. Together, we worked to sharpen their message and identify more companies that might sign on, doubling down on our outreach to utilities and unions.

As we neared the end of June, I felt worn down. It was one thing to read a suspenseful novel and quite another to live it. At that moment, I wanted to skip to the end of the story and find out how it all turned out: whether I would end the year with healthy babies, whether our clean electricity program would survive, whether Congress would ever pass a climate law. But there is no fast-forward button in life. I had to live through every long hour.

12

INFRASTRUCTURE WEEK

Just after noon on June 24, a gaggle of reporters waited at "the sticks," a spot on the White House driveway where impromptu press conferences were often held. They hoped to catch the bipartisan group of senators after a meeting with the president. To their surprise, Biden walked out. Wearing a red-and-blue-striped tie and flanked by ten senators, he announced they'd struck a deal. On his left, Democratic Senator Kyrsten Sinema from Arizona was beaming in a bright red-and-white dress. On his right, Republican Senator Rob Portman from Ohio sported a pastel blue tie. They had all dressed for the occasion, decked out in bipartisan colors, as if clothing alone could hold the deal together. But press conferences are not a bill on the president's desk. The agreement was a measly two pages. Ron Klain, Biden's chief of staff, knew he still had a lot of work to do.

Before he could even meet with these senators to talk next steps, the deal almost fell apart. Two hours later, the president held another press conference in the East Room of the White House. He stood at a podium, in front of heavy yellow curtains, and opened his briefing book. A 224-year-old portrait of George Washington, saved before the White House burned in 1814, watched over the proceedings. Ron was buoyant. As a White House veteran of three separate administrations, he knew that very few days in the West Wing were this successful.

During the half-hour briefing, a journalist asked the president what he'd do if the bipartisan deal reached his desk before the reconciliation package. "I'm not signing. Real simple," Biden replied. "The bipartisan bill, from the very beginning, it was understood there was going to have to be a second part of it. I'm not just signing the bipartisan bill and forgetting about the rest." This was what the progressives wanted him to say. But by the time the president left the room and boarded a helicopter, Ron's good day had gone bad. The Republican senators were pissed. Why would they spend their time working on a bill with the president if he wouldn't even say he'd sign it?

Ron didn't believe that holding up the infrastructure deal would provide leverage over Manchin on reconciliation. This was the Sinema-Portman deal, not the Manchin-Portman deal. Manchin just didn't care as much about the infrastructure bill. On top of that, Ron knew that Congress was bad at multitasking. The longer the infrastructure negotiations dragged out, the longer it would take to pass a climate bill. But progressives in the House, and activists on the outside like me, disagreed.

Ron knew the White House had to release a statement. That was the only way to put the infrastructure deal back together and get Portman to stop yelling at him on the phone. Within 48 hours, he'd crafted a press release clarifying that Biden believed both bills were equal priorities, and the president would sign whatever reached his desk. Everyone was equally annoyed and assuaged, and the deal was back on.

With the bipartisan deal making progress, more senators wanted to get involved. They'd been frustrated by the endless talk and jokes about infrastructure week during Trump's first four years in office and were eager to join. The group swelled to 22 senators. They met around an enormous table, alongside Brian Deese and Steve Ricchetti from the White House. Steve, a longtime lobbyist who served in the Clinton White House and later worked for Biden during the Obama administration, was there to handle the politics. It was Brian's job to nail down the policy details. This was proving challenging.

Brian wanted to increase climate and environmental justice spending in the package. And he refused to add anything that increased pollution. But the Republicans were firm that they wouldn't touch any policy with the word *climate* in it. Brian argued that cutting pollution could be bipartisan. Clean air and water shouldn't be controversial. The Republicans agreed, in principle. But they remained stubbornly opposed to including anything on climate. So Brian spent his time contorting policies to make them fit into the bill without a whiff of the word.

He asked the Climate Policy Office staff, including Sonia and Jahi, to help find clean energy and environmental justice investments on the president's wish list that could be rebranded. Sonia didn't agree with the strategy. Before she came to the White House, she'd spent years trying to pass bipartisan climate policy, having grown up surrounded by Republicans in rural Ohio. She'd believed that they could be convinced to act on climate with enough effort. But in the end, she concluded she'd failed. "We wasted at least a decade trying that," she said. She worried they'd lose another decade to the same dumb strategy.

But that decision was above her pay grade. All she could do was try to make the deal as good as possible. She worried that anything that wasn't included might never make it into law. And she worried that any climate or environmental justice investments stuffed into the infrastructure bill would be mangled by Republicans to the point of futility. Deciding whether to include a policy or leave it out was a tightrope walk.

Sonia's top priority was getting the clean electricity plan through Congress. After Gina told the press that the program might not get included in the reconciliation package, Sonia called to ask if she really meant it. Gina responded that of course she supported the plan and wanted to see it in the deal. She just didn't know if it would happen. Being in White House leadership meant that comments could spiral out of control and lose context. Sonia felt reassured, and redoubled her efforts to get the program passed into law. If she could get it into the bipartisan deal, that would be the best outcome.

But she was running into headwinds, even internally. For the past several months, she'd led an interagency working group to develop the program. The staff at the Department of Energy were particularly helpful, running models of how a tweak in the payment or penalty would affect clean energy growth. But an economist from the Treasury Department was far less helpful. She joined most calls with a sour look on her face, declaring that the plan wouldn't work and refusing to propose solutions. Sonia suspected this economist had even organized outside academics to start attacking the policy. And Sonia knew that this economist had called Manchin's office to complain about the program, breaching protocol.

If they couldn't even get moderate Democrats on board, Sonia knew that it wouldn't work for Republicans either. Eventually, she figured it would not fit in the infrastructure deal, and it would be better to wait for the reconciliation package. They could redesign the program to appeal to Manchin, who held the pivotal fiftieth vote and could singlehandedly sink the entire bill. The problem was that electric utilities remained noncommittal about supporting the program, even with $150 billion in federal funding to make it work. Sonia kept trying new strategies to unstick the negotiation. Her latest idea was getting a group of utility executives a meeting with the president. If they were invited, they would have to come. So she wrote a memo, pitching Ron and other senior White House leaders on the idea. But it went nowhere. Maybe Energy Secretary Jennifer Granholm could meet with the utilities instead. But that didn't happen either. So far, this effort was a dry hole.

Feeling stalled, Sonia decided to refocus on transmission policy. Big power lines were not being built because no one could agree on who was responsible for paying for their construction. Over the past decade, new transmission had fallen by 80 percent. And this was a problem because reaching higher shares of clean power required a vast network of high-voltage lines. Without this transmission backbone, clean energy projects would be stuck waiting years for permission to connect to the grid. Sonia knew there needed to be a clear formula to split the transmission costs between developers building clean energy and customers benefiting from it. She wanted to include language in the bipartisan law to give the Federal Energy Regulatory Commission clear authority to figure this out.

Sonia went to Manchin's staff on the Energy Committee, armed with her most persuasive one-pager. But their response wasn't exactly positive. So she came up with a slightly different variation, only to get the same response. Then she did it again. And again. Sonia had a way of trying to force a horse to drink. One of her senior colleagues, who had worked in the Obama White House, pulled her aside to politely inform her it wasn't going to happen. Manchin's team was smart. Her creative new language wasn't going to trick them or make them change their minds. It was time to stop trying.

Instead, the infrastructure bill added funding for carbon capture and sequestration, direct air capture, and hydrogen. There were billions of dollars proposed to set up hubs and demonstration projects for these technologies. Sonia worried these programs would be poorly designed and lead to more pollution. If hydrogen production wasn't done right, companies could pull dirty energy off the grid and use it in electrolyzers, claiming the fuel was clean when it wasn't. There was also $6 billion proposed to save the nuclear fleet from shutting down due to low electricity prices. That was important, but if utilities got a nuclear bailout in this bill, they'd be less inclined to negotiate over the bigger clean energy program in reconciliation. Sonia didn't understand why the White House had included these things that Manchin wanted in the bipartisan bill rather than holding onto them for the reconciliation package. If he got everything he wanted in this first bill, why would he do another?

Meanwhile, Jahi was busy working with the White House Environmental Justice Advisory Council. It was set up in the first week of the Biden administration to find ways to support communities burdened by pollution. For decades, grassroots environmental justice activists had pushed for federal funding to clean up pollution from lead pipes, ports, and power plants. The communities most burdened were usually poorer and majority Black, Hispanic, or Indigenous. When he joined the White House, Jahi knew he wanted to deliver for these communities above all else. This was easier said than done, and he had to figure out how to make it happen. To make progress, Jahi was helping develop a tool that could identify these polluted communities to direct funding where it was most needed. This aligned with the president's Justice40 Initiative, which aimed to direct 40 percent of the benefits of climate and clean energy investments to disadvantaged communities.

But to spend money in these communities meant they needed to have money to spend. Building cleaner infrastructure should clearly fit within the bipartisan bill, if the White House could just convince some Republicans to go along with it. Vice President Kamala Harris was particularly passionate about electric school buses, and thought this might be a bipartisan issue. When she was still a senator, she introduced legislation to fund them. Now she wanted to get this spending into the infrastructure package.

There was precedent for the idea. Across the country, environmental justice advocates were pushing for the same thing at the state level.

Cinthia Moore grew up in East Las Vegas, a poorer Hispanic community in Nevada. As a kid, she spent hours on the school bus every day. As soon as her son was old enough to start riding the school bus, he began having headaches and trouble breathing. Cinthia thought back on her own chronic migraines from childhood, and realized the diesel fumes might also be making her child sick. Working with other activists, she successfully pressured the Nevada legislature to get ready for federal funding for electric school buses. Now they just had to get that funding into the infrastructure package. And with VP Harris pushing from inside the White House, it felt like it could happen.

Environmental justice advocates were also pushing for a national green bank. This would offer low-cost financing to clean energy projects in places that struggled to access capital—speeding up the transition and helping lower-income communities participate. Too often, wealthier communities adopted solar or electric vehicles first, because tax credits were the policy that drove clean tech, and poorer people didn't have much tax liability. A national green bank could expand access to clean technologies. Jahi wanted to help make that happen.

Inspired by Obama, Jahi had worked as a community organizer in DC in 2008, when he was just out of college. He remembered talking to young men who looked just like him, sitting around in job training centers, wanting to work but frustrated with the limitations a criminal record imposed. Meanwhile, DC was struggling and needed more workers. It was being hit regularly with hundred-year storms, and the old sewers couldn't handle the extra water. People living in Bloomingdale, just south of Howard University, were waking up to eight inches of water in their basement. Jahi thought these folks looking for work could help solve this problem. He ran a campaign to create an Infrastructure Academy in DC, which would train folks in clean jobs, like constructing green roofs and rain gardens to hold excess water.

In a meeting with a city manager, Jahi laid out his idea. But the manager was flippant. He explained that Jahi simply didn't understand how bond financing worked. Even after setting up the Academy and proving this skeptic wrong, Jahi kept thinking back on that meeting: he was smarter than that guy. He would teach himself how bond financing worked. And that's how he stumbled on a bigger idea: green banks.

State and local institutions were already using public financing to build clean infrastructure and create green jobs. Jahi knew that if this idea went national, it would help so many communities. With his program in DC, he'd seen it with his own eyes.

Like Jahi, everyone else in the White House climate office wanted to get a green bank into law. As Brian worked through the infrastructure negotiations, he tried to get the Republicans to agree to some of these environmental justice priorities, including the green bank, support for electric school buses and funding to remove lead pipes. Brian asked Jahi to develop these ideas further, in case he could fit them into the infrastructure deal. Jahi was excited about the prospect of getting these policies that he'd dreamed about for years enacted into law.

But when Brian discussed the green bank with the Republican senators, their response was bizarre. They wanted to build a technology-neutral green bank. It made no sense. Fossil fuels did not need more financing—conventional banks were doing more than enough. This version of a green bank wasn't what the Biden administration was trying to build. They would end up with a greenhouse gas reduction fund that didn't reduce greenhouse gases. Brian told Jahi it would be better to try again for the green bank in reconciliation. Maybe they could still get some funding for electric school buses and lead pipes in this infrastructure deal.

13

$3.5 TRILLION

After Senator Bernie Sanders put out his $6 trillion spending proposal in the middle of June, his committee began finalizing the budget resolution. This would give the rest of the committees instructions, including spending limits. Schumer's team wanted to ensure that the package would pass with 50 votes on the Senate floor. To do that, Manchin would need to engage—and after six months of stalling, he continued to show disinterest. They'd also need to persuade Senator Sinema, another moderate Democrat who was holding out.

As Schumer's policy director, it was Gerry Petrella's job to figure out how to get from 48 to 50. He thought it through for a few days, then pitched Schumer on using the Budget Committee to try to make progress. The way Gerry saw it, it was a microcosm of the broader Democratic caucus. It included the chairs of other key committees that would negotiate the bill: Ron Wyden on finance, Patty Murray on health, and Debbie Stabenow on agriculture. These were senior leaders who would do the actual work negotiating everything from paid leave to farm subsidies and clean energy tax credits. Plus, the Budget Committee had moderates like Mark Warner, who could serve as a substitute for Manchin and Sinema. "If you can get Sanders and Warner to align, then we've got a shot," Gerry told Schumer. It was either a brilliant or a terrible idea, but Schumer decided they should try it.

Even though he oversaw all policy, climate was Gerry's white whale. After working in the Senate for a dozen years, he'd watched almost nothing happen on the file, despite the urgency. As a staffer, he knew that passing a climate bill would be his mic-drop moment. Maybe if he did that, he'd be ready to leave his job.

Gerry's focus on climate sharpened when he became a father in the summer of 2018. His wife also worked on Schumer's policy team, and together they tried to balance difficult jobs with parenthood. In 2020, their two-year-old son started to miss milestones. It was the height of the pandemic, and finding support was next to impossible. They were both working endless hours, trying to pass a bill to dig the country out of a massive economic hole. Being isolated, exhausted, and unsure what was wrong with his son was the hardest thing Gerry had ever experienced. Eventually, his son received an autism diagnosis. But Gerry struggled to accept it. He wondered if he had done something to cause this: the stress of his job, the long hours at work, the loneliness of the pandemic. As the months passed, he began to lean on his Catholic faith from childhood, starting to accept that this was just the way God had made his son. His child was special.

But Gerry also knew that a warming world would be particularly hard for kids like his, who struggled with a disability. On hot days, his son had a hard time. And he knew that as the crisis accelerated, people with special needs could be overlooked when disasters like Hurricane Sandy hit. He knew the best way to protect his kid was getting this damn climate bill done.

The day after the senators returned from the Fourth of July break, Schumer called an in-person meeting with the Democrats on the Budget Committee. It was Gerry's job to prepare the agenda, and Adrian's job to handle the climate file. Schumer aimed to get the senators to agree to a framework, locking in the top line number. He knew a group of 22 bipartisan senators were also meeting, just down the hall, to try to hammer out an infrastructure deal over the next 48 hours. Schumer wanted to beat them to it.

By the time the Budget Committee got started, it was already evening. The senators, their staff, Brian, and other White House leaders crammed into a ceremonial room in the northwest corner of the Capitol, close to the Senate chamber. The gilded mirror, vaulted ceilings, and elaborately tiled floor added to the importance of the occasion. They'd assembled in this room several times before, but tonight felt different. They could almost smell an agreement in the air.

Senator Mark Warner was particularly vocal. He wanted to ensure the package was paid for with revenues from a tighter tax code and cheaper federal prescription drugs. To make it all work, Warner proposed cutting the package by about half, down to $3.5 trillion. But the proposed cuts did not focus on climate: the senators agreed those programs needed to stay, including the clean electricity program at $150 billion.

Instead, healthcare funding was under threat, much to the dismay of Sanders, who wanted to deliver dental, vision, and hearing coverage for older Americans. He was also fixated on keeping $10 billion in solar funding for lower-income households. As the hours passed, Sanders grew more and more frustrated. He could not understand why they were leaving so many programs to help everyday Americans on the cutting-room floor.

It was near 9:00 p.m. before most of the issues were settled. Sanders, speaking up in his role as chair, asked Schumer to step into the hallway, their staff trailing behind. When the door shut, Sanders turned to Schumer and asked him to make a promise. He had conceded on so many priorities as he watched the committee whittle down his package. He said he would agree to this much smaller number but needed assurance that they would not keep hacking away at progress. "Chuck: You've got to promise me that this is it—this is what we're going to do."

Gerry had seen this move many times. Senators always wanted a guarantee from their leader. But Schumer couldn't make any promises that night—he wasn't Manchin or Sinema, who were withholding their votes. "Bernie, I can't promise you anything," Schumer said as they stood in the hallway. "This is what we need to do to move this process forward. You've got to do this for the country, for the party." Sanders grudgingly accepted that this was the best he would get. The conversation was over.

When the two men reentered the room, it was clear the negotiations were coming to a close. Before wrapping up, Schumer wanted to ensure they were all on the same page, so he asked everyone to verbally confirm their agreement. Sensing that history was being made, everyone seized on the opportunity to speak. As Gerry and Adrian listened to senator after senator, they wished they could record the moment.

It was Senator Stabenow from Michigan who spoke most eloquently. After more than two decades in office, she'd seen so many missed opportunities to advance the Democrats' agenda: climate action, but also paid leave for new parents and tax benefits for the middle class. Like everyone else, she'd had her list of things she wanted to accomplish, including support for electric vehicles. But Stabenow could also see the bigger picture. And that night, she was beaming. If they could get this deal passed, it would be the largest expansion of the federal safety net in over 50 years. They had been working on this package for decades. Looking around the room at her colleagues, she saw the people she'd fought alongside. And they had finally done it. "I'm just so proud to be here with all of you today," Gerry heard her say, her voice becoming emotional. "We are making history."

A gaggle of reporters waited just down the hallway, hoping to catch the Democrats and Republicans on their way out of these rival late-night meetings. Gerry and the communications staffers thought they should capitalize on their progress and hold an impromptu press conference in the Ohio Clock corridor, announcing to the world that Democrats had struck a deal before the bipartisan group. Schumer, never shy of a camera, turned to his fellow senators to organize who would stand where, who would speak, and what they would say. They wanted to present a unified front.

After a few minutes of discussion, Schumer flung open the door and began marching down the hallway, with Gerry and Adrian at his side. When they reached the press huddle, the senators arranged themselves like a kindergarten class photo. Chuck Schumer stood in the middle, with Bernie Sanders and Mark Warner on either side, serving as visual reminders of the range of views from left to right in the caucus. Farther out were Patty Murray and Debbie Stabenow, still beaming. Schumer began his speech, explaining that the deal they'd struck was fully paid for and would accomplish all the goals President Biden had laid out. After he was done, he let the other senators address the press, then fielded a few questions.

On their way back to the office that night, Gerry and Adrian felt like they were trailing clouds of glory. Still: the workday wasn't over. Back at the office, they watched as Schumer called more senators. He wanted to let them know about their progress and enlist their help in selling the package to Manchin. The next day, Manchin's Energy Committee would be marking up the infrastructure bill. Schumer needed Democrats to build goodwill with Manchin by voting for it. By the time Adrian and Gerry got home close to midnight, their kids were fast asleep. They wouldn't stay long, because President Biden was joining the Democratic caucus the next day for lunch to talk about progress on both the reconciliation package and the infrastructure deal.

When Adrian arrived back at work early in the morning, Gerry called him into his office. After working for Schumer for years, Gerry had trained himself to speed-read several newspapers first thing in the morning. Schumer did that when he woke up, and Gerry knew he could be grilled at any moment. That morning, when Gerry opened his copy of *The Washington Post*'s July 14 edition, he skimmed the story about the deal. But that wasn't what Gerry wanted to show Adrian: it was the photo. The photographer had captured Schumer just as he left the negotiation room, with Adrian and Gerry flanking him—his literal right-hand men. They smiled and laughed, amazed that they'd made the paper and proud of their progress on this impossible quest. Adrian walked to the lobby's stack of newspapers and picked up his own copy. When he got home, he cut out the photo and stuck it in a frame to display in his office. He wanted to remember this feeling.

14

THE WAITING ROOM

That same morning that Gerry and Adrian read the newspaper, I got my own good news. On July 14, Senator Tina Smith announced our clean electricity plan was officially in the budget deal at a whopping $150 billion. Ricketts and I were thrilled. We'd been working toward this day for over a year and spent the morning celebrating with colleagues. Forgetting the time, I glanced at my watch and realized I was running late for yet another doctor's appointment—now that I was 26 weeks pregnant, there were weekly scans. On our way out the door, my husband and I locked eyes. "You know what this means," he said. I read his mind: if the policy had momentum, the babies couldn't possibly be doing well. Normally scientific, we'd become superstitious.

We arrived at the high-risk clinic and headed back to the exam room. Dr. Rask soon walked in, his usual chatty self, full of off-color jokes. He sat down and began scanning my belly for the tenth time. Now his best guess was Baby A weighed just one pound six ounces. Like *The Very Hungry Caterpillar*, I ate and ate and ate. But most of it stuck to me: I was in the 99th percentile for weight gain, while Baby A was in the 1st. No matter what I did, the smaller baby wouldn't grow. Twenty minutes later, when he'd finished the scan, Dr. Rask skipped his usual rant about the climate hoax and grew serious. "I've got some bad news," he said. Baby A could take a turn for the worse at any moment. If these troubling signs continued, I'd need to be scanned three times a day. We could carry on with outpatient monitoring for now. But pretty soon, that would no longer cut it. I needed to be ready to go to the hospital at a moment's notice.

It was devastating news. The odds of the babies being born healthy had fallen. When we got home, I started to prepare: asking colleagues to take meetings in my place, getting my finances in order, and trying to set up some kind of nursery. Now 18 months into the pandemic, many people were vaccinated, resuming their lives, and enjoying summer. My reality could not have been more different—I was headed to the hospital and would only leave after my life was radically altered. The best-case scenario was a one-month stay, an early but safe birth, and then two babies in neonatal care for months. The worst case was a very short hospital stay and the death of one or both twins. There were no "good" outcomes—at least not in the short term.

When I returned for my next scan four days later, Dr. Rask called the whole thing off. The troubling signs were gone. I felt relieved yet wary. He said I could continue outpatient monitoring twice a week and come back first thing in the morning on July 22, three days later.

Increasingly, I was struggling to sleep. And it wasn't just the stressful pregnancy or my round belly growing more uncomfortable by the day. I was worried Congress was going to fail to act on climate again. When President Biden announced the infrastructure framework the month before, on June 24, Ricketts and I pored through the details and realized there was very little climate spending in the package. Sure, the deal included worthy investments in public transit, clean water, and transmission lines. But it would hardly cut carbon pollution.

As if the Earth itself was responding to Congress's inaction, an unprecedented heat wave hit the Pacific Northwest the day after the deal was announced. Seattle topped 108°F and Portland hit 116°F, both breaking records. Roads buckled and cables melted. A small town in British Columbia called Lytton recorded the highest temperature Canada had ever seen. And the next day, that town burned to the ground. Watching the news unfold, I knew people would die. That's what happened when temperatures hit record levels and most homes had no air-conditioning. That's what happened when fires hit. Hence the trouble sleeping.

As the heat wave raged, Sunrise staged a protest in front of the White House with Representative Ocasio-Cortez and other congressional leaders. More than 500 young people showed up to demand Congress and the president enact meaningful climate policy that summer. Like me, they were frustrated by an infrastructure deal that wouldn't cut carbon pollution. After the rally, they blocked all entrances to the White House, angering chief of staff Ron Klain, who felt he was working his ass off to deliver on climate. But pressure was growing. The next day, more than 100 labor leaders were on the National Mall calling for bold climate action.

In response, the White House issued a memo from Sonia's boss, Gina McCarthy, the national climate advisor. Ricketts and I read it closely like it was a magic eye and a hidden meaning would emerge if we stared long enough. Gina said that while the bipartisan infrastructure bill contained important investments, it wasn't sufficient. She called on Congress to pass a climate bill with clean energy tax credits, a Clean Electricity Standard, and the Civilian Climate Corps. It was a worthy list, but we needed even more: investments in clean buildings, transportation, and industry. I believed the White House remained committed to climate action. But Congress? Exactly how much were they prepared to invest in our future? That's what I wanted to know.

The day after this memo dropped, the international nonprofit Greenpeace released videos they had covertly filmed with two ExxonMobil lobbyists. Posing as job recruiters, they asked the industry insiders to explain their strategies. The recordings revealed the oil giant was working behind the scenes to ensure the infrastructure bill excluded climate spending, sending its CEO to meet with senators. They said they met with Manchin's office weekly.

And then, as if the Earth wanted to add an exclamation point at the end of a dramatic week, a fossil gas pipeline burst in the Gulf of Mexico, lighting the ocean on fire. As it grew, it resembled an eye, like a god watching Congress dither, disgusted.

As June turned to July, the climate disasters continued, with hurricanes and wildfires and drought. And then the budget resolution was finally announced in the middle of the month. While we were thrilled that CEPP had made it, Ricketts and I had advocated for a range of clean electricity policies: tax credits, funding to help rural electric co-ops retire coal plants, investments in air pollution enforcement, support for fossil fuel workers, and transmission funding. We wanted to figure out if these investments were included. But it was hard to tell which programs made it into the reconciliation framework, because the Budget Committee only released top-line figures.

We began calling around to try to understand what was hidden inside each committee's instructions. Manchin's Energy Committee had been allocated nearly $200 billion, while the Environment Committee only got $67 billion. This gap would prove consequential months later, but at the time, I didn't give it much thought—I was so focused on ensuring funding for the CEPP.

Although Ricketts and I weren't spending as much time on transportation funding, we knew the climate bill needed to include more. The bipartisan framework proposed new EV charging investments, no doubt because Brian knew how badly they were needed. But this wasn't enough. We also needed to rapidly scale up EV adoption through consumer incentives. These had existed for years but were capped for each car company—and most had blown through their caps. To meet Biden's climate goals, a majority of car sales had to be electric by 2030. At the time, just one in 16 were. Rapid growth wouldn't happen without new incentives, particularly for lower-income Americans. This was the problem Adrian was working on most closely for Schumer. He told us that he'd managed to get a vague direction for the Finance Committee to fund "transportation tax incentives." But Manchin hadn't signed off yet. And Adrian worried he never would.

Working with Ari Matusiak and others at Rewiring America, we were continuing to push for home electrification funding. The building sector was particularly depressing: pollution had grown over the last 15 years. Together, we came up with another list of policies to make progress: manufacturing grants for heat pumps, support for states and cities to adopt stronger building codes, and funding for green retrofits in affordable housing. But we spent the most time on a new rebate program for home electrification with Senator Heinrich. We wanted to secure as much spending as possible to help Americans put heat pumps in their homes, with the funding targeted at low-income and middle-class households. And we wanted to ensure that federal support for gas furnaces stopped. They were already cheaper than heat pumps.

Efficiency advocates remained stubbornly opposed to cutting fossil fuel subsidies and didn't seem to care whether rebates went to the people who needed them the most. The problem was, Schumer's team insisted we all cooperate. If there were going to be two rebate programs, we would need to divide up a pot of money assigned to Manchin's committee. It wasn't ideal.

Attempting to work out our differences, Ari and I ended up on a call that summer with the efficiency groups and their bill sponsors, Senator Chris Van Hollen from Maryland and Representative Peter Welch from Vermont. Our electrification champions, Senator Heinrich and Representative Castor, also joined. Heinrich, an engineer by training, argued intelligently that the technologies were ready for full electrification. And Castor made an impassioned plea to cut fossil fuel subsidies for the sake of our grandchildren. The other side insisted that gas furnaces were better for poor folks and people of color—a justification I found maddening. To me, it seemed like we clearly had the stronger argument, but the efficiency groups couldn't hear what we were saying. The conversation ended in a stalemate.

Frustrated and sleep-deprived, I went back to the doctor for another scan on July 22. It did not go well. The flow was reversed in Baby A's cord, meaning she could go into distress at any time. I needed to go to the hospital. I arrived later that day, assigned to a large private room with a window overlooking the street below. I was not allowed to leave this room, not even to walk the halls. The doctors hoped to keep me for a month. With each additional day I remained pregnant, the babies had a higher chance of being born healthy. The goal was getting to 28 weeks. And then 30. And ideally 32. And heck, why not 34. But that first day, I was just shy of 27 weeks, and things were not looking good.

Living in the hospital was easier than expected. I didn't worry as much because we checked in on the babies three times a day. I simply filled the time between scans with phone calls and meetings from my hospital bed.

A week in, I woke before sunrise. It was the last Friday in July, and the infrastructure deal was advancing. I'd just written an op-ed for *The Atlantic* explaining that this bill was not a substitute for climate action. This article landed me an opportunity to go on the television show *Democracy Now!* I said yes, even though I'd be secretly calling in from the hospital at 5:00 a.m.

The problem was, the sun wouldn't be up that early. I asked Edith, one of the nurses who'd been taking care of me, if she had anything I could use to light up a corner of my room. An hour later, she reappeared, pushing a giant surgery lamp. It was the sort of thing doctors use when extracting a baby from a birth canal. Technically, we weren't misusing hospital equipment because one of its bulbs was broken. It still cast a beam bright enough to mimic a blazing sun.

When my husband arrived a few hours later, we laughed at the absurdity of the scene. I made him take a picture of me in my makeshift TV studio, a giant surgery lamp leaning over me, just to remember the moment. Those are the last photographs of me pregnant.

15

A ONE-PAGE AGREEMENT

Just two weeks after Schumer secured the $3.5 trillion deal with the Budget Committee, Manchin marched into his office late in the day. For the past two weeks, Schumer had been trying to get Manchin to guarantee he'd support the reconciliation package, if Schumer moved the infrastructure deal forward. When Manchin arrived at his office unannounced on the last Wednesday in July, he held a one-page document with conditions for his support. Manchin didn't want to begin negotiating the package until at least October. And he would only spend $1.5 trillion—less than half of what the other senators had just agreed to. He insisted that Schumer sign his name on the document, right then and there. He'd even prepared a space for his signature.

Schumer demurred, trying to navigate the strange situation with grace. He told Manchin he'd work in good faith to try to meet his requirements. But Manchin didn't want good faith. He wanted Schumer to sign on the dotted line. The two senators were close: Schumer had helped Manchin win his first race back in 2010. And now they were arguing like brothers.

Manchin believed that Schumer was going too far left. He felt he'd pushed him too hard in the spring, when the Senate passed the pandemic rescue package. He was constantly talking about how they'd burned him on that bill. Manchin now believed they'd spent too much money and given too many handouts. He'd become obsessed with the national debt, requiring an aide give him an update on the total every morning. He would not repeat the same mistake. This time Schumer would have to do what he wanted, not the other way around.

Eventually, Schumer relented, but he told Manchin he was going to try to persuade him not to do some of this stuff—and he was going to write that below his signature. Schumer also asked Manchin to keep the document a secret, saying "Don't tell anybody." Manchin agreed.

When Adrian and Gerry saw the agreement afterward, they were baffled. They'd never seen anything like this. Manchin had forced the majority leader to sign a vague document, with no prior warning. This wasn't the way senators negotiated.

The document had five points on climate, which Adrian read hastily. It said Manchin wanted his committee in charge of the clean energy standard. Adrian took this as a good sign, because it implied Manchin wanted that policy in the package. The other points focused on supporting fossil fuels. This wasn't surprising. And that was it. Adrian was thrilled: he knew they could deliver on this. He told Schumer that when it came to the climate demands, this was a great deal. They should take it and run.

But not everyone would share Adrian's assessment. Two months later, Manchin would send the document to Speaker Pelosi while she sat watching the Congressional Baseball Game. When she read the secret deal, she was furious. Pelosi immediately called Schumer, demanding to know why he hadn't shared Manchin's topline. Schumer defended himself, pointing to below his signature where he had added in scratchy handwriting, "I will try to dissuade JOE on many of these." It was his attempt to claim he hadn't made a deal.

16

BIRTHDAY

The day after my early-morning TV interview, I slept in. It was Saturday and work had slowed, leaving fewer things to distract me from the scary situation I was in. A staffer in Representative Welch's office called to talk about the efficiency rebates. We were trying to get them to change the draft legislation and phase out funding for gas appliances, since they refused to stop funding them altogether. But they wouldn't commit to doing it for six years, if they did it at all. They were constantly using Manchin as their excuse, but the truth was, the energy efficiency coalition was also going to Manchin to push for these subsidies. By the afternoon, there was a new proposal to split the funding between the electrification and efficiency programs. Other than that, it was a slow Saturday.

That day, Edith was working again. A serious nurse in her sixties, she'd struck me as stern when we first met. But as the days passed, I grew to appreciate her experience and skill. "It's a quiet day," I told her when she entered that morning. Edith did not appreciate my offhand remark. "You used the Q-word," she told me. I didn't understand what she meant. "What's the Q-word?"

"Quiet. You should never say it's quiet."

Like many nurses, Edith was superstitious. Bad things tended to come in threes. When the moon was full, the maternity wing would fill up. And if someone uttered the word *quiet*, it would inevitably bring chaos. I laughed, dismissing her magical thinking. It was just a slow day.

After lunch, I spent an hour on the phone with a friend working in the Biden admin, then fell asleep, missing the appointed scan time. Two hours before shift change, Edith woke me. "I let you sleep as long as I could," she said, as she prepped the stress test, eager to get it done before she clocked out. Edith strapped the monitors across my giant belly and tried to catch the babies' heartbeats. I watched the rhythmic lines pop up on the screen for the thirtieth time.

But that afternoon, Baby B would not cooperate. She was punching and kicking the monitor, disrupting the data. Edith kept repositioning the device, trying to get a clean fifteen-minute observation. We never watched the babies for this long. As an hour passed, tiny Baby A, tucked in my pelvis, was still. We watched her heart rate crash one, two, then three times: a clear sign of stress. Edith was worried. "I don't like what I'm seeing," she said. My husband grew quiet. When Edith's supervisor came by to remind her she hadn't taken her break, she refused. Instead, she called Dr. Rask, insisting he come to the hospital immediately.

Dr. Rask arrived within the hour. He wheeled over an ultrasound machine, sat down beside my bed, and started yet another test. This time he looked closely at the baby's breathing, movement, muscle tone, heart rate, and amniotic fluid. When he finished the exam, he said that Baby A had scored just two out of ten possible points. I knew what this meant.

"I think we're going to have to deliver them," I said, scared.

"Wait! I need a minute to think," he replied. "I want to make the right call." This was the most important decision Dr. Rask had to make in his job. Ten seconds passed, and then he said, "Yes, we have to deliver them."

Half an hour later, Dr. Rask was pushing me in a wheelchair toward the operating room. During stressful moments, I turned to humor to cope. "I've decided on my birth plan," I said, setting up my punch line as we moved through the wide hallway. The doctors and nurses were always complaining about couples' overly complex "birth plans," so I knew he'd worry about what I'd say next. "I have two requests," I told him. "First, don't kill me. And second, don't kill the babies." Dr. Rask laughed, and said he'd try his best.

I wouldn't be choosing the position to deliver in, what music would be playing, or who would be present. Dr. Rask had decided to fully sedate me, in an attempt to protect the babies, and as a result my husband would not be there. Still, more than a dozen people were in the room: doctors and nurses packed into the space, ready to attend to both me and the newborns. As I lay down on the operating table, I continued cracking jokes, trying to distract myself from the terror. It annoyed some of the doctors who were just trying to get on with their work, but it helped me cope. Then the anesthesiologist sedated me, and I was gone.

I woke up groggy and disoriented, unsure of where I was or how much time had passed. My glasses were missing, so my husband put prescription sunglasses on my face while I looped on the same questions: "Are the babies alive? How much time has passed? What time is it now?" As the sedatives wore off and the opiates kicked in, the answers kept slipping away.

An hour later, a doctor from the NICU arrived to update us. Dr. Olatokunbo had curly gray hair pulled back in a bun and a complex accent that was hard to place—we'd later learn she spoke four languages. Her bedside manner was kind but vigilant. She told us that the babies were both alive and doing well. But there had been some complications. Baby B's transition was not ideal. She had needed to be resuscitated and was now on a ventilator. We had no idea what that meant. I began to cry, and my blood pressure shot through the roof. Alarms started beeping.

"I think I've upset you," Dr. Olatokunbo said, concerned. "We can talk more later. I think I will go now."

When I finally stabilized, it was close to midnight. Unable to walk, I was wheeled on a gurney through the long hallways. We passed through the NICU's security doors and entered the space for the first time. The soundscape shifted: beeps and alarms filled the air. My husband was directed to stop at the soap station and scrub all the way up to his forearms. Then they parked me in the center room next to two incubators, where six people working the night shift attended my two newborns. From a distance of several feet, I strained to see the babies, both encased in incubators with clear plastic shells. Born at 29 weeks—two and a half months early—they were tiny, fragile beings the size of small eggplants. Baby A weighed not much more than a pound and a half.

Days passed before I could touch them. A week or more until I could hold them. My husband and I knew we would be coming back to the hospital for months.

17

THE NICU

We spent the next few nights in the hospital as I healed from the birth. I was covered in strange bruises and struggled to get out of bed. Each day, my husband wheeled me down the long hallways and through the heavy double doors to the NICU. After stopping at the washing station to sanitize, we would enter the middle room with its half a dozen patients, and peer at our babies sleeping in their incubators, unable to touch them. That oxytocin rush that so many mothers experience during childbirth, bonding them to their children and bringing in their milk, never came. Instead, I was dissociated from what was happening. If motherhood was a neighborhood, I was lost several blocks away.

On our first night back home, I held my husband's hand as we lay in bed and said we should share the highlight of our day. It was a spontaneous ritual, designed to keep us sane, and we continued it as the days passed.

When we returned to the NICU each morning, time felt distorted, both faster and slower. There wasn't much to do, so we mostly sat around. It was almost like meditation, except without the silence. Monitors were attached to each baby to detect changes in heart rate, breathing, and oxygen. When the system noticed that any of these metrics were outside the normal range, an alarm would sound, alerting the nurses. Premature infants do not know how to breathe, so they often stop. The nurses called this a "desat" because the baby's blood was desaturated of oxygen. Other times, the child's heart rate would also crash, an event they called a "brady," a shorthand for bradycardia. Either of these events would trigger the alarms, and the nurses would rush over to stimulate the baby, trying to remind the tiny being to breathe. To avoid these events, my daughters were hooked up to various breathing machines, which blew air into their noses continuously. Swaddled beneath white blankets and covered with equipment and lights, my husband thought they resembled little robots. Despite their size, they were fierce beings, constantly ripping out the wires and tubes from their bodies, triggering the alarms. The nurses had to replace their medical devices again and again.

But even with all this support, we watched our babies desat regularly in that first week. With around 20 babies in the NICU, an alarm was almost always going off. We would glance up, look to see if it was one of our children, then feel guilty if it wasn't. When we got home each evening, I could still hear the beeping. The sounds were phantoms, ringing in my mind.

It was a week before I was allowed to hold each baby. With all the wires and tubes, it took a team of three people just to get one of them into my arms. The nurses marked the milestone with a small, colorful flag that hung by each child's bedside. For "micropreemies"—children born weighing less than 1,000 grams, or 2.2 pounds—the nurses made sure to celebrate even the smallest progress because they knew the families would be there for months. When they were taken off a ventilator, they got a flag. When they turned one week old, they got another. The nurses were trying to tell us that every little step counted.

There was one big event that we were waiting for that the nurses did not mark with any flag: the results from the babies' brain scans. When I was pregnant, Dr. Rask explained that if we did not catch distress in utero fast enough, the fetuses could suffer small seizures that left behind brain bleeds. There were degrees of severity, and in many cases the brain would heal. But we would want to know if it had happened. It was one of the earliest indicators of their future health.

In those first days, the attending NICU doctor came by regularly to give us an update. Dr. Olatokunbo, who was in the room during the birth, was often in charge. A cautious doctor, she cared deeply about making the right calls for her tiny patients. One afternoon during the first week, she stopped by. After rattling through a long list of updates, she said there was one other thing she meant to mention before leaving: their brain scans came back clean. Neither baby had any brain bleeds. We were thrilled. "You buried the lede there, Dr. Olatokunbo," I said, laughing.

As the weeks passed, we fell into a rhythm, shuttling between our home and the hospital. Every 12 hours the babies were assigned a nurse, so most days we had a new personality to navigate. They would help us transfer the babies to our chests, where we would cradle them for a few hours as they slept. On some days, I would squeeze in a few work calls from a storage room packed with medical equipment. And then we went home for dinner where I multitasked: pumping milk while writing whatever policy document was needed. Or dealing with our insurance company, which had been sending endless notices that the medical care for my children was "unnecessary." We'd return again in the evening, hurrying back to the hospital before visiting hours closed. At night, I slept like the dead. And then my alarm would go off in the pitch black, reminding me it was time to get up and pump. Every other part of my life was pruned away. It was work, then hospital, then work, then pump, then work, then hospital, then sleep, then pump, then sleep, then repeat. Somewhere in there, I found time to eat.

And then this pattern abruptly stopped. A few weeks after her birth, my smaller baby began to have trouble breathing. After one night shift, the nurses called in the morning to tell us she'd had 12 desat events overnight. They decided to increase her breathing support and keep her in the isolette until she stabilized. Hours turned to days. I couldn't hold her for a week. With the stress, my already meager milk supply stopped altogether. It was a scary time. Eventually, she stabilized and started to breathe again without long pauses.

To explain my increasingly frequent absence from policy calls, I started to tell a few more people I was working with that I'd given birth to twins. They were always surprised. Spending so many hours in the hospital, I had to rely on other people like Ricketts to carry on the work without me. He told me the bill text for the CEPP was being negotiated between Senator Smith's office and the House Energy and Commerce Committee staff. They planned to run the draft by the parliamentarian to get a read on whether it would comply with the Byrd Rule. And the team at Rewiring America was still trying to nail down funding for the electrification rebate program.

During these first few weeks of my babies' lives, it became clear how interdependent my life was. There were limits to what I could control.

We say "I gave birth" to a baby, a book, an idea. But it's never really true. There are always others holding us in a delicate net. I could see clearly that my kids would not be alive without all the doctors and nurses. Or the people who invented the incubators and the protocols for threading PICC lines through their tiny veins. I could see that these fragile policies wouldn't be alive either without a network of people who believed in them. Who kept pushing, even when things looked bad.

By mid-August, our odds of passing a climate bill felt as low as ever. The bipartisan infrastructure bill, devoid of any big climate investments, was now scheduled for a vote in the Senate. It was almost certain to pass. Ricketts heard the progressives in the House were facing pressure from Speaker Pelosi to vote for it next. We worried this might prove a fatal misstep. If the infrastructure deal became law, Manchin might never vote for a climate bill.

18

45 PERCENT

By late July, Brian knew they needed to pick up the pace to finish the infrastructure bill before the August recess. After days of meeting at tables big enough to accommodate 22 senators, he realized they weren't going to make progress sticking with the same strategy. It had worked well up until this point, but now they were going around in circles. To finish the deal, they had to shrink the table. The Democrats gave the White House their proxy, and the Republicans chose Senator Portman.

For the next several days, Brian worked from Portman's office, heading to the Hill early in the morning. By then, Brian had grown to respect him. A tall, thin man with neatly parted silver hair, Portman had served in Congress for more than two decades. He was a pragmatic, old-school Republican. As the hours wore on and day turned to night, his staff ordered round after round of takeout to keep Brian in the room just a little longer. Often, he wouldn't be home with his family till after 9:00 p.m. And the next day, he returned to do it all over again.

As they neared the end of the process, Brian surprisingly found common ground with the Republicans on environmental justice investments. He'd managed to convince them to add $55 billion for clean drinking water, which would remove lead pipes concentrated in communities of color and poorer neighborhoods. They also agreed to make the largest public transit investment in history, with billions for rail and clean buses. Plus, thanks to environmental justice advocates and VP Harris, the package also included $5 billion for electric school buses.

Finding revenue proved to be the stumbling block to finishing the deal. Everyone wanted to be able to say that they were not adding to the national debt. But the Republicans could not for the life of them come up with ways to raise money. Brian offered one idea after another: increasing the corporate minimum tax, closing loopholes, a carbon border adjustment fee. But the Republicans didn't have the votes for those policies. The revenue sources they finally came up with were questionable at best. They proposed repurposing unspent COVID funds, delaying a new Medicare rule, and using proceeds from a telecoms auction. When Brian tried to grasp these revenue sources, they slipped through his fingers like smoke. Like many Republican bills before, the infrastructure deal would increase the deficit—by more than $250 billion.

But that ended up not being a problem. On Tuesday, August 10, the Infrastructure Investment and Jobs Act passed the Senate by 69 to 30. Schumer scheduled the vote on the budget resolution immediately afterward, and it similarly passed. Republicans railed against the reconciliation bill. They claimed it was "reckless spending" that would increase the deficit, even though it was fully paid for—unlike the bipartisan deal.

Now the House just had to vote on the infrastructure bill. But Representative Jayapal's progressive caucus insisted they would wait until they had some kind of deal on the reconciliation package with its social policies and climate investments.

Watching the infrastructure package move forward with so little funding for climate freaked Sonia out. She wanted the reconciliation package to catch up. And she knew the only way she could help make that happen was with Schumer's team. But the White House had a strict protocol. The Office of Legislative Affairs functioned as an air traffic controller for overly ambitious rank-and-file staff like Sonia. They couldn't just fly around the Hill talking to whoever they liked. It would cause chaos.

Sonia wanted to work with Tim Ryder in Schumer's office. Sonia had met Tim years earlier, when she was still at Energy Innovation and he was just starting to build the ambition tracker with Adrian. She was immediately impressed: in a decade working with folks on the Hill, she'd never met a staffer who understood carbon pollution, sector by sector, in so much detail. He seemed eager to meet and talk through the specifics. Sonia couldn't quite tell if he was just a nerdy intern or if his work would actually go somewhere, but she figured it was worth making the time.

Years later, Tim's ambition tracker was fully operational, and he was deep into quantifying the pollution cuts from the draft package. Schumer's model was very detailed. Tim wanted to be thorough so that Schumer could precisely estimate both the climate and infrastructure packages and show the world they were meeting the moment. He would spend hours trying to quantify how many tons could be saved from investing in a single train.

Sonia knew there was no point in having the White House replicate the Senate's work. She decided to ask a colleague in Legislative Affairs if she could meet with Tim to work on the ambition tracker together. He agreed but said he'd need to come, too, so his office could keep tabs on all Hill outreach. She warned him these meetings were likely to be long and technical—some people might even call them tedious. But that didn't dissuade him. He said he'd be happy to join.

The call started in the evening, after most of Tim's colleagues had clocked out and Sonia's son was in bed. The Legislative Affairs guy joined, but he just listened. They quickly dispensed with the sectors Tim had already nailed down. Some areas, like electricity, were easier to estimate. Looking at the reports, their best guess was that the CEPP and the tax credits combined would deliver a 19 percentage point reduction below 2005 levels by 2030.

But when it came to things like agriculture, forest conservation, and wetlands, it was much harder to make an educated guess. They needed to figure out how much stimulating sea kelp growth would increase carbon sequestration. Sonia reminded Tim of a recent paper from the National Academy of Sciences, and he pulled it up. The only trouble was, kelp grew in myriad places and at many different rates. Tim knew they would need to dig into the footnotes. After ten minutes of reading, Sonia chimed in that they could assume it was eight tons of carbon sequestered per year for every hectare of protected wetlands. But Tim also needed to know how much it would cost per acre. They'd have to figure that out later.

And then there was the matter of scale. How many acres of sea-kelp-growing wetlands were there, exactly? Tim suggested pulling up some maps to check out the mangrove extent in the Southeast Florida region. He reminded Sonia that if kelp growth accelerated, marine mammals might make a comeback—had she seen these studies showing how whale carcasses sunk to the bottom of the ocean? Oh, yes, she had—those whales were full of carbon! They would need to include that in their estimate too. Half an hour into the call, their minder from Legislative Affairs was still listening, but his eyes had glazed over. He told Sonia that he trusted her to take it from here and logged off.

Now that they had permission, Sonia and Tim continued to hold these calls every week or so. They were long, often stretching six hours, but they never found them boring. They would go over the latest version of the package, reviewing their methodology for quantifying the tons and the cost, and make updates based on new data.

As the weeks passed, Sonia started to multitask, pulling out the Halloween outfit she was making her son, even though the day was still two months away. It was her favorite holiday, and she wanted to be prepared. Her son had just started walking, so she was sewing him an elaborate owl costume with long wings, each feather stitched by hand. As she worked away, Tim meticulously archived every fact they found, just in case an enterprising journalist ever came to ask tough questions about their estimates. He wanted to have the receipts ready. They'd grown to love these meetings, each finding a fellow nerd who cared about the smallest of details. There was no one else in the world like them.

After weeks of effort, Tim and Sonia were adding up the numbers, trying to guess how much progress the infrastructure deal plus the draft reconciliation package could deliver. By mid-August, President Biden himself was asking to see their preliminary results.

It was right around then that Gerry came up with an idea: Schumer should write a letter to his colleagues in the Senate, updating them on the latest estimates from the ambition tracker. House progressives had received the infrastructure bill poorly, believing its investments in highways and roads would only increase pollution. Plus the reconciliation package included dozens of climate programs, making it tricky for senators to wrap their heads around what was even in it. Adrian and Tim had built their model for exactly this moment.

Tim got to work, making a colorful pie chart that broke down the pollution cuts, program by program in both bills. In the accompanying letter, he called out the CEPP and tax credits specifically, to remind members that these policies delivered the biggest progress. When Schumer looked over the draft letter, he was pleased. He'd told his team numerous times that they needed to stay near President Biden's goal of cutting pollution by 50 percent that decade. Tim's analysis showed they could hit 45 percent—close enough that executive action and state leadership could plausibly fill in the gap. But, he reminded his team for the tenth time, they couldn't slip much farther. "No matter what," Schumer said, "we cannot fall below 40 percent."

Then one morning in early September, Tim called. Sonia was trying to squeeze in an errand on her way to work, getting passport photos for her son. She answered just as she pulled into the parking lot. He said he was letting her know, as a friend, that he didn't think the CEPP was going to make it into the final bill. It just wasn't looking like Manchin would go for it. Sonia pushed back, telling Tim this must be bad intel. If it fell out, they would lose the biggest pollution cut in the package. He said he could be wrong, it was just an educated guess. Still, when Sonia hung up, she started to cry. Now she was late for work, and there wasn't enough time for the photos. She'd have to get them another day.

For the next month, Sonia told herself that Tim was mistaken, even as all signs pointed in the opposite direction. She began to have long phone calls with utility executives and economists, who unloaded their list of complaints about the policy. And then there were press reports that Manchin was hanging out with some of the utility CEOs that Sonia knew opposed the program.

She was frustrated that they hadn't made more progress with Manchin. Her boss, Gina McCarthy, had been working closely for months with his wife, Gayle Manchin, since she became co-chair of the Appalachian Regional Commission. It was a federal institution set up under the Kennedy administration that supported economic growth across West Virginia and the surrounding states. Everyone said Gayle was a smart, kind person. But whatever positive relationship the White House held with his wife, it didn't seem to extend to the man himself.

For months, Sonia had argued that the president should make time in his wall-to-wall schedule to meet with the utilities. No one could make a better case for the program than the president. And if the utilities got on board, maybe that would unstick Manchin. But despite her persuasive memos, the meeting never got scheduled. She grew paranoid that White House leadership was throwing the CEPP under the bus. Maybe they never really believed in it. Maybe they always saw it as a sacrificial lamb.

Getting desperate, she started working with the Department of Energy to come up with versions of the program that Manchin and the utilities might accept. They tried one where the funding was centered on reliability. Another with higher grant payments. Then a third that dropped the penalty altogether. It felt like no matter what they proposed, the response was lukewarm at best.

If anything, the utilities seemed to be hardening against the idea. One day, Sonia's boss called to tell her that she needed to lay off the utilities a bit. Someone had called from the Edison Electric Institute, the private utility association, to complain about Sonia. They said she was "harassing" them. She found this an absurd characterization. All she did was ask them to be specific. She wanted to know what the White House could do to get them to yes. "I don't care what it is you think you need," she had told them. "I just want you to tell me something." Their response was silence.

The utilities fundamentally misunderstood Sonia. Colleagues and outside groups saw her as a very kind person. And she was effective. This combination led Jahi to nickname her "the velvet fist." Had the utilities provided any ideas, she would have done everything she could to make it work. And she would have done it fast.

Sonia was so productive that she often finished her work in half the time as others. Occasionally, working moms would commiserate with her about how much they were missing out on being with their kids, and she would try to relate. But the truth was, she spent lots of time in the evenings with her son. He was just learning to talk, his first sounds babbling like a bumblebee. Sonia loved to record him when they took walks around their suburban rental. Together, they stared up at the night sky as she taught him the words for everything they saw: moon, stars, tree, house, car. He tried to repeat them back to her, and she laughed.

Late that September, he had his first birthday. Since he was just starting to eat solid food, she made his cake from chia seeds and apricots. It turned out disgusting. But they celebrated anyway, laughing at the terrible cake, her little family of three. They were happy.

Right around this time, her husband called one day while she was at work. He was frantic. The house they were renting was surrounded by giant trees, and as the summer waned, there were dramatic, windy thunderstorms. That afternoon, as their son was napping in their bed, a giant branch fell through the roof, a foot away from where he slept. And now it was raining in their bedroom. Her husband was freaked out, telling her he wasn't sure how much longer they should live there. "We need to call the owners to tell them to maintain these damn trees!" he said. Sonia tried to calm him down, but when she hung up, she was spooked. If this was an omen, it couldn't be good.

19

TESTIMONY

By September, my daughters were still in the hospital, but they were doing well. As they stabilized, I went back to working longer hours on the climate bill. The electrification rebate program was still being negotiated. The current plan was to split $20 billion with the efficiency program, which would fund gas furnaces for wealthier Americans. At Rewiring America, Ari Matusiak and I were frustrated. We weren't making progress in eliminating fossil fuel subsidies, and our funding kept shrinking.

Meanwhile, the clean electricity program was on shaky ground. On paper, it looked like we were making progress. There was now bill text and a hearing scheduled in the House Energy Committee. But the policy's detractors were growing. Someone would say, "Did you hear what the parliamentarian said in that meeting?" or, "I heard Manchin won't vote for that," and before we knew it, everyone was worked up. I started to get calls from journalists with vague concerns about how the CEPP might undermine grid reliability, or the parliamentarian might not like its design. Ricketts and I had grown used to quieter, internal disagreements within the environmental movement. But these fights were becoming public. We joked that it felt like a video game: as we progressed through the levels, the bosses were becoming harder to defeat.

Economists, both inside the Treasury Department and outside government, honed in on one particular issue: "resource shuffling." They worried that utilities might just move clean electricity credits around, rather than building new energy projects. In their minds, this gaming of the system would break the program. CEPP would give grants to utilities that increased their clean power at the necessary pace, tentatively set at 4 percentage points a year. If they missed their goal, there were penalties, but they were much lower than the grants. This asymmetry between the carrot and the stick was a political compromise: there was no way Congress would heavily penalize utilities. Doing that could even cause electricity bills to increase, when the goal was lowering costs. But these economists worried that the lower penalty might create perverse incentives. Utilities that couldn't meet the target might just sell their clean electricity credits to utilities that were closer to the goal. And people were buying into this critique. Ricketts and I heard that these economists had talked three Democratic senators into holding some kind of hearing on the topic.

Ricketts knew one way to calm folks down: write a memo. He started yet another document and sent it over. Together, we explained why these worries were unfounded. The program included significant reporting and auditing requirements. Utilities caught gaming could be subject to criminal prosecution, civil penalties, and the requirement to return grant dollars. And the vast majority of utilities were heavily regulated at the state level. CEPP would force these utilities to report their grants to state oversight bodies. With a requirement that clean energy grow year after year for every utility, we argued it was unlikely that they would just cheat the system and absorb penalties.

The bigger problem in our minds was that they might not move fast enough, leaving too few credits to go around. But this could be addressed by reducing the requirement to below 4 percentage points a year. Ricketts and I expected that to happen as negotiation over the program continued.

As our critics grew, we began to realize at least some of them were bad faith. One day Jesse Jenkins, a professor at Princeton, reached out. We'd known each other since grad school at MIT and had been working together on the CEPP. He had a newborn plus a toddler, and was trying to balance being an academic, a policy analyst, and a dad. That day, he showed me this Twitter account he'd found. I pulled it up and saw its entire purpose was criticizing the policy. It would reply to every single thing Jesse, Ricketts, or I posted about it, like a stalker. The account claimed we were "corrupt" because we'd helped fund a study estimating potential job creation from CEPP. We'd disclosed this fact—it wasn't a secret or anything problematic. But their goal was manufacturing controversy. The account also linked to a newly created website, which attacked the program. We all thought it must be funded by the fossil fuel industry. They would lose out if our program became law.

But these were small problems. Ricketts and I knew that, ultimately, we had an audience of one: Senator Manchin. And, like Tim had told Sonia, we'd started to hear that his tenuous support was shrinking. On September 12, Manchin went on the Sunday talk shows, including *Meet the Press*. And he did not hold back. After rattling off some statistics about the grid mix over the past two decades, Manchin concluded: "The transition's happening. And that clean energy standard, they want to spend billions of dollars to have utilities do what they're already doing. Makes no sense to me to pay a utility to do what they're going to do anyway." Then he repeated the same talking points on CNN, adding that he was also worried about reliability. Ricketts and I watched with dread. This was not good. When CEPP passed the House committee two days later, it was hard to celebrate. If Manchin wasn't on board, it didn't matter what the House thought.

At the time, I was preparing to travel to DC. Senator Heinrich was hosting a hearing on electrification and asked me to be an expert witness, alongside Ari from Rewiring America. It was hard to imagine leaving my children in the hospital, even for a few days. But it was harder to pass up a chance to speak on their behalf, as the youngest members of our human family. I decided I could make a very short trip.

A few days before I was scheduled to leave, I woke up early and started yet another long day of calls. We had meetings with a utility that was worried about how its nuclear fleet would qualify under the CEPP. After that, I briefed a House committee, and then checked in with Senator Smith's office about the bill. My written testimony was due that day, so I was also trying to finish it. The work day stretched longer than expected, and my husband grew anxious. We needed to get to the NICU.

We arrived at the hospital in the afternoon. Walking to our children's bedside, we saw their nurse was clad in a plastic gown and gloves. This had never happened before. My heart sank. Dr. Olatokunbo was working that day and came over, explaining that one of my daughters had caught a cold. Out of caution, a nurse had sent some swabs to be cultured several days earlier. And one of these had tested positive for MRSA, a hospital superbug. The protocol in cases like these was to isolate the patient. My daughter barely weighed three pounds. Putting her in isolation seemed extreme, particularly since it meant she was back on a feeding tube, after making progress eating from a bottle. We were devastated. I contemplated canceling the trip.

But as I looked at the facts, I became convinced that my daughter didn't even have this superbug. They had cultured both children's eyes and noses. Yet the culture only came back positive in one child's eyes. Why wouldn't the theoretically infected baby also have it in her nose, mere millimeters away? And why didn't my other daughter, attended by all the same nurses, also have it? Nothing made sense. Given how common this superbug is throughout hospitals, I started to suspect the sample was contaminated on its way to the lab. I asked the doctor to run the test again to check for a false positive. She agreed it was a good idea, but we both knew it would take days. And in the meantime, my baby was stuck in isolation.

Since we had two patients in the hospital, the doctors worried we might infect the uninfected baby. They required us to gown and glove as well. This meant we couldn't hold our sick daughter skin-to-skin, which was critical for her development. I understood they were trying to prevent an outbreak spreading to even more babies in the NICU. But the costs for my child were quite steep.

At a certain point, being in the hospital can become more harmful than helpful. Whether it's a newborn or an elderly person, there are risks in staying too long. We'd reached that point. My husband and I realized that if we could get our other daughter home, at least we would be allowed to hold our child who was stuck in the hospital.

But the doctors were hesitant to discharge my healthy daughter, even though she'd just graduated from her isolette to a bed. I insisted she would be ready soon. Looking at the calendar, I told Dr. Olatokunbo I would bring her home in one week's time, the day after returning from my trip. We all agreed to work toward this deadline.

On the plane ride to DC, I finalized my remarks and decided I would talk about my newborns. Very few people knew I'd been pregnant. Since everything was remote, I was able to hide my high-risk situation from all but the closest collaborators. There were hundreds of people I'd worked with who I knew would watch my testimony and find out I'd just delivered twins. After hiding it for so long, I was nervous about sharing.

The morning after I landed, I walked to the Capitol, crisscrossing the city's many parks full of statues of men. Wearing professional shoes for the first time in over a year was painful, as they dug into my ankles. My work clothes were too tight, my body larger than it used to be. Usually a confident public speaker, I felt anxious. I was riding the hormonal waves of new motherhood and worried about maintaining my composure when I talked about my kids, who were both still in the hospital. Plus, I would be addressing some politicians who were hostile to my message, people like Senator Ted Cruz from Texas.

When I reached the Hart Senate Building, I stood in line and passed through the security checkpoint. After a few wrong turns, I found my way to the hearing room. Together with the other witnesses, we waited in a small holding area, mingling with the Democratic members. Senator Heinrich shared that Manchin's office, located just next door to his, was full of photos of Manchin with famous people like Bill Gates. We wondered aloud if that might be a way to try to convince him to pass the climate bill.

At the appointed time, we walked into the hearing room. The ceilings were wood-paneled and the central wall was clad with white marble, making the space glow. It was where Ruth Bader Ginsburg and Sonia Sotomayor faced tough questioning before their confirmations to the Supreme Court.

I sat down at my assigned spot behind the name card that read "Dr. Stokes" and waited. I would have just five minutes for my remarks. When my turn came, I told the assembled senators and representatives that the American people were watching them. For over three decades, scientists like me had warned Congress that climate change posed a dire risk. Those warnings had come to pass, with disasters worsening every year. And still Congress had failed to act. I urged them to end this gridlock and pass a climate bill this year, while we still had a chance.

With a minute left, I turned to my conclusion. "I'd like to close by sharing something personal," I began, my voice quavering. "I have two newborn babies that are two months old at home right now, and I have left them to come here and speak with all of you because I know the urgency of this moment. I'm here not just to speak on behalf of my newborn children but on behalf of all the children in the United States. I'm also here to speak on behalf of young people who are deeply fearful about what their future holds. And quite frankly, I'm here to speak on behalf of all of us, because we are already seeing what climate change is doing to the places and the people we cherish most. This is a pivotal moment in history. The world is watching. Either Congress will pass a bold climate investment package this fall or we will lose the last best opportunity we have and waste another decade. We don't have any more decades left to waste to act on the climate crisis."

When I invoked my young children, I felt the energy in the room shift. I could see the politicians soften—even the Republicans. Saul Griffith, from Rewiring America, was sitting behind me, and he started to cry. Children remind us of our common humanity. They remind us of what's at stake if we fail.

When I returned to Santa Barbara the next day, one of my babies was healthy enough to come home, as I'd predicted a week earlier. That afternoon, when we laid her in her bassinet for the first time, she could look up at those same colorful flags that hung above her incubator. She'd just received her final one. It read "I graduated from the NICU."

But we still had another child in the hospital. Thankfully that weekend, the second culture came back negative, and she was let out of isolation. Over the next several weeks, my husband and I shuttled back and forth to the hospital, trading off so one of us could be with each baby at any given moment.

20

BEGINNINGS AND ENDINGS

In early October, Secretary Jennifer Granholm left her office in the Brutalist Department of Energy building and headed for the Hill. She was on her way to meet with Senator Manchin. Now in her early sixties, Granholm still retained her youthful looks, and wore her blond hair short. Her staff had come to see her as the literal embodiment of her agency: full of energy. When her senior staff first met her, she told them to expect impatience. She wanted her team to move fast.

Granholm and Manchin had known each other for years, serving together as governors in the early 2000s. He was a fan and often called her "kiddo" rather than "secretary"—a habit that drove her staff nuts. But this lack of decorum rolled right off her back. She wanted to be close to the senator, and a nickname was a good sign. As was the fact that he always took her phone calls.

For the past several months, Granholm had met with Manchin and Senator Tina Smith, trying to make a deal on the clean electricity program. Each meeting was more hopeful than the last, with Manchin telling them that he wanted to get it done. But he remained averse to finalizing a specific plan. Even his staff seemed unsure what their boss thought. In an attempt to get Manchin to nail down the details, Granholm proposed a meeting with him and a group of utility executives. He agreed, and they set the date for early October.

It did not get off to the best start. Manchin arrived late. He was coming from the Oval Office, where he'd spoken with the president about the broader reconciliation bill, and their meeting had run long. Granholm, undeterred, opened the conversation in full persuasion mode. She told Manchin, the assembled utility executives, and the Energy Department staff that she was open to compromise, but they needed to get this clean electricity program done.

The utilities listened attentively. By then, some of them were interested in receiving billions of dollars to do the right thing. Exelon, which had a large nuclear fleet it struggled to operate, thought the program might help keep these plants open. And PSEG from New Jersey was also supportive—it had recently committed to 100 percent carbon-free power by 2030. Still, some utilities wanted changes to Senator Smith's framework. Southern Company said the requirement should be slashed to just 2 percentage points of extra clean power every year. Duke Energy and American Electric Power argued the penalty should be eliminated, or at least cut in half. The Energy Department staff listening into the call felt optimistic that the utilities were open to compromise. Maybe they could land a deal at 3 percentage points of new clean power a year, with a $20 penalty for every megawatt-hour the utilities fell short. Granholm thought this was workable.

But Manchin was distracted, only partially listening. After six months of effort, Granholm's staff found themselves frustrated by how vague the senator remained. As the meeting wrapped up, Granholm proposed that Manchin's team sit down with Energy Department staff and the utilities to hash out the details and then bring a final proposal back to the senator.

I knew this meeting had happened, and heard some reports that it had not gone to plan. But I was focused on getting my second child home from the hospital. Although she had almost tripled in size since birth, she still didn't have enough fat to stabilize her temperature. That was one of the NICU's requirements for release.

Every few days, I tried to set a deadline with the nurses for her discharge. But that day would come and go, and she would still be there. The front-desk workers started asking when it might happen. I would tell them Friday, then Sunday, then Tuesday. And that day would come and go, and she would still be there.

Ricketts's daughter had just been born healthy. I was thrilled for him. But now he was out of the office for at least a few days. That's when we heard a rumor from Speaker Pelosi's office that they were being told to cut $160 billion in climate spending from the package. At the time, it totaled

$670 billion, so this would slash climate spending by almost a quarter. And the strange thing was, this directive didn't come from Manchin. It came from Arizona Senator Kyrsten Sinema.

Sinema had publicly raised concerns with the deficit, but if this was her position on climate, she was wrongheaded. We couldn't nickel and dime the climate crisis. Just one year of climate disasters could cost $160 billion. I asked one of the centrist organizations we were collaborating with to talk to her office and report back. They said her staff complained about the cost of the CEPP. By our estimate, it wasn't expensive: just $22 per ton of pollution saved. But if Sinema's staff was complaining about spending, it confirmed the speaker's office report.

Even though Ricketts was on paternity leave, he never wanted to miss anything. I called him, and he agreed it was a good idea to plant the story. After that, I reached out to Coral Davenport at *The New York Times* and told her what I'd heard. She said Sinema was an enigma who rarely talked to the press, and no one knew exactly where she stood on the climate package. So this was an interesting scoop. She would try to confirm it.

A few days later, on Friday, October 8, the story hit like a bomb. Sinema's office was given a chance to comment and denied trying to shrink the package. But with Arizona so vulnerable to extreme heat waves and drought, she was heavily criticized for pushing cuts. That evening, people kept sending me the article, and I tried to act surprised. Maybe this bad press would keep her in line in the future.

Later that night, as I was just sitting down for dinner, Sonia called. I had been trying to reach her to hear her take on the utility meeting with Granholm and Manchin. But that wasn't why she was calling. It was late on the East Coast, and she was out walking her dog in the dark. Like Tim had said to her six weeks earlier, Sonia was calling me as a friend. Things were not looking good for the CEPP. For weeks, she'd done everything she could to try to get Manchin on board. But they were negotiating with a weather vane. Even if Manchin seemed open to an idea one moment, he would change his mind the next time he was asked. After the meeting with Granholm and the utilities, his staff had cancelled the follow-up, saying it was never going to happen. And rumor had it he'd decided to run for reelection in 2024. Or maybe he'd met with fossil fuel lobbyists after the meeting and simply changed his mind. People said he had a tendency to repeat whatever he'd last heard.

Getting this policy passed felt both inevitable and impossible. It was the most consequential one in the package. The Biden administration was behind it. Democratic leadership in both chambers of Congress backed it too. Even the utilities were slowly getting on board. But Manchin had hardened against it. And he had the final say.

Sonia told me she had been having a recurring dream all week. The bill had passed, and she was out at a bar. Everyone else was celebrating, but she was crying. "I think you're going to need to call up a dream interpreter for that one," I told her. We laughed. But we were both devastated. She thanked me for my work on the policy. And then she stepped in dog shit. The universe wasn't even trying to be metaphorical.

I hung up, shell-shocked. But there was no time to process the call: it was 7:30 p.m. I needed to pump and rush to the NICU before the front desk closed at eight. If I was even two minutes late, the security guard treated me like an intruder. That had happened a couple nights before, when I'd arrived at the hospital just as Sonia called and stayed too long in the parking lot talking to her. They almost didn't let me see my daughter. I couldn't let that happen again.

I got to the hospital with a few minutes to spare and took the elevator up to the NICU. When I sat down across from my daughter's isolette, and the nurse transferred her to my chest, the sadness hit. It was one of those recurring nights, when we thought it might be her last evening in the NICU. Even though I desperately wanted her home, there was sorrow too. It felt like things were ending.

But the next morning, my daughter's temperature dropped again. She wouldn't be coming home after all. And the CEPP wasn't dead—not yet. In Ricketts's absence, another close collaborator gave me a pep talk, and I decided to spend my Saturday doing whatever I could to try to save it. I called progressive members of the House and the Senate, reminding them that the CEPP was the most consequential policy in the package. And then I called the centrist groups, who had relationships with Sinema and Manchin, to strategize.

But by Monday, rumors were swirling. People texted that they heard the CEPP was dead. Who was this coming from? Someone said it was a senior person in the White House. Others said it was a staffer on the

Senate Environment Committee. But no one could say definitively. And the fact was, it wasn't true. Not yet, anyway. I kept thinking of Mark Twain. When a reporter called him up after a story was published saying he had died, he replied, "Reports of my death are greatly exaggerated."

But in DC, rumors morph into reality. Whatever is repeated enough can become fact. If everyone believed the CEPP was dead, it wouldn't be long before it was. We generally worked a week ahead of the news landing in the paper. When these rumors started, I knew it was only a matter of time.

By Wednesday, we were closing in on my due date, and one of my daughters was still stuck in the hospital. When I called the NICU that morning to see how she was doing, the nurse replied that she'd gained more than six ounces over the last 24 hours. She'd weighed her again and again, then used a different scale. But no matter what she tried, she got the same result. This was a huge jump: in the past week, the baby had barely gained any weight. The news emboldened me. She needed to come home now.

Dr. Olatokunbo wasn't on duty that day, so when we arrived later that morning, we met with another doctor and made our case. We would do whatever was necessary: check her temperature every three hours, keep the house at 80°F, weigh her daily. With my mom visiting, we had more adults than babies. And, I added, my other daughter was already doing so much better at home.

The doctor agreed to call our pediatrician, and if they were on board, then they would release her. Initially, we thought she'd come home that day—a result my husband had pegged at 0 percent odds that morning. But we realized she needed one more vaccine. So we agreed to keep her there just one more night. She would come home 75 days after her birth, just before her due date. Somehow, it felt like success.

As soon as we made the decision, Dr. Rask showed up. We hadn't seen him in almost three months, since our first days in the NICU. It was like bumping into a high school classmate who disappeared after graduation, yet here he was a decade later in the grocery store produce aisle. Dr. Rask said he'd followed the babies' charts and was pleased at how well they were doing. In his non-politically-correct way, he assured me they didn't have cerebral palsy. I took his sudden appearance as confirmation that it was time to go.

The next morning, I pulled the dozen colorful flags off the wall, and we said our goodbyes. I was not sad to leave the beeping monitors behind, but I knew I'd miss the people. The doctors and nurses and respiratory therapists and cleaners and folks at the front desk. Working together, they'd performed a miracle.

We walked our daughter out the main entrance feeling like we'd masterminded a jailbreak. It was her first time outside since she was born, and the air was smoky. A wildfire was burning just up the coast. We loaded her into the car, and I played "Hey Jude" by the Beatles—it was what I sang to her during those long weeks in the hospital. Now she could hear the original rendition for the first time.

We'd barely made it a block from the hospital—the song had not even reached its crescendo—when my phone began to blow up. Journalists were hearing reports that CEPP wasn't looking good. Rumor was, the White House was poised to release a new framework for the bill without our program. When we got home, I sat on the couch for half an hour, holding both my daughters and marveling at just how identical they looked. Then I got back to work, spending hours on the phone with journalists, trying to reassure them about CEPP's prospects—even as I'd started to suspect it had a terminal illness.

The next day I woke up in a good mood. It was my due date, and both of my children were now home. We'd made it happen—we were finally a family, together under one roof.

That morning, I reached Keane Bhatt, who worked for Representative Jayapal in the House Congressional Progressive Caucus. We'd been talking for the last few months about the policy. I reminded Keane that CEPP delivered the biggest pollution cuts in the package. We needed the progressive members to stay firm. Keane told me he understood the stakes and would do his best to keep the caucus united on this.

Later that day, Ricketts and I had a meeting with EEI, the private utility trade association. Before it began, he told me we needed to push them hard. But barely two minutes into the call, the EEI staffer cut me off: "Let me stop you right there—we have tried everything." She said she had sold the policy to utility CEOs, kept them in line, and come up with endless solutions for Manchin's daily concerns. Manchin just wouldn't go for it.

The man was a brick wall. After we hung up, Ricketts called me back in a bad mood. He demanded to know why I didn't push them harder, like we'd planned. I told him there was no point.

And that's when Coral Davenport reached out. We'd talked just one week earlier, when I'd given her that scoop on Sinema. But I knew that wasn't why she was calling. She told me she had the story, that she knew Manchin had killed the CEPP. I couldn't help but think this was some kind of karma for planting that Sinema article. Unsure of what to say, I asked for five minutes to think before giving a comment, and hung up.

Sonia told me it was time to accept we'd lost—she was farther along in the grieving process. But a staffer from Senator Smith's office, who led on the policy, told me to deny it. When I reached Ricketts on the phone, we tried to make sense of this conflicting guidance from our closest collaborators. He wondered if there was still more we could do. But we couldn't deal with that now: I was on a deadline. Together, we worked out what I would say, word for word. "This is the most important climate policy in the package. We need the CEPP to meet our climate goals." Ricketts agreed this was the right tone to strike.

When I called Coral back, the quote came out differently. I stuck to my lines but couldn't hold back emotion. I added that without it, we couldn't meet our climate goals, concluding, "This is pretty sad."

Within minutes of the story going live, Adrian sent my quote back to me over text. It was his way of saying he wasn't happy. "We're working to recover the lost emissions," he wrote. "You didn't leave much room for optimism."

"Sorry," I replied. "I was trying for a better quote."

"It's okay," Adrian responded. "People are still processing the shittiness of the situation." He was already working to fill the hole CEPP had left. "We've got a bunch of stuff cooking," he wrote. "We're crunching numbers to figure out options but if you have any good ideas, send them our way."

When I woke up the next morning, Adrian's message was lodged in my mind like a mantra. There were still things we could do. It didn't matter that it was Saturday or that I had two newborns at home. I needed to get back to work.

21

THE RACE TO SAVE TONS

Inspired by Adrian's late-night texts, I began Saturday morning searching for ways we could fill the hole CEPP left behind. Based on modeling from a number of groups, it was somewhere between 350 to 400 MMTs annually by 2030. That was a large hole.

In the afternoon, I walked to the corner store to buy a copy of *The New York Times*. My quote had made the front page of the October 16 edition, right above the fold. Normally, this was something to be proud of, but it felt embarrassing. I'd made the cover because I'd failed. I texted an image of the paper to Ricketts, and he called, bummed out. When I told him I was working on ways to make up for the lost tons, he struggled to get on board. "Maybe we should be a little humble," he said. We were the ones with the big idea that blew up in our faces. The most painful thing was we were so close: 49 senators would have voted for it.

Then Ricketts began a postmortem, analyzing all the ways we might have run the play differently. He wondered if the package should have included a carbon tax, so that Manchin could have killed that instead. But there *were* senators pushing for a tax, and it never even made it into the framework. He wondered if the policy should have been routed through the Environment Committee, which Senator Tom Carper from Delaware chaired. But Manchin was on the record, in that letter he'd forced Schumer to sign in July, insisting it go through his Energy Committee. The truth was, we had to cross through the treacherous territory of Manchinland eventually. It was either in his committee or at the eleventh hour on the Senate floor where Manchin could veto anything.

To make ourselves feel better, Ricketts and I told each other that our efforts might have saved other policies. Maybe we had kept Manchin busy, training his fire on us. There was talk that he was complaining about the clean energy tax credits to the billionaire Bill Gates. Maybe we had helped protect those tax credits, and the buildings and electric vehicles policies.

After Ricketts wrapped up his assessment, I told him I planned to keep working on filling the hole. It was hard to psych myself up when I knew I could fail again. But the alternative was a planet on fire. I told Ricketts I would keep him updated on what I found.

The best remedy for this mixture of grief and embarrassment was work. Sitting at my desk in the corner of my bedroom that also doubled as a pumping station, I started a new document. I reminded myself of the silly phrase I'd come up with for moments like these: "The spreadsheet is mightier than the sword." If I could get the data and organize it, that would help us make progress. I was determined to find new ways to cut pollution.

One possibility was having the federal government pay people directly for solar and heat pump installations. These "direct pay" tax credits allowed everyone to access federal funding, including low-income people who didn't owe the government much in taxes. A wide coalition of solar companies, climate organizations, and environmental justice groups had been pushing this for years. Ricketts and I agreed with the approach and promoted it every chance we could. But there were two problems: one, staff on the Senate Finance Committee didn't like it; and two, Manchin might not either. Regardless, I put it on the list, adding that we should also stop federal funding for fossil fuel appliances.

Then I called Sonia to ask for her ideas. She was thinking about ways to make more progress in the electricity sector. Her boss, Gina McCarthy, understood that losing the CEPP could undermine activists' support for the whole package. She'd encouraged Sonia to find new ideas to cut pollution while giving activists a political win.

Sonia and I both knew tax credits alone wouldn't clean up the electricity system fast enough. Utilities needed funding to retire debt on dirty fossil fuel plants and shut them down. Rural electric cooperatives were particularly burdened by coal debt. The package already contained $2 billion for the Energy Department to help utilities refinance at lower rates and retire their debt. Sonia thought if this program was expanded, more utilities could benefit. In my spreadsheet, I proposed increasing it to $10 billion.

While the package contained other funding for rural electric co-ops, Sonia worried about municipal utilities in urban areas. They also needed low-cost financing to retire dirty assets and build clean infrastructure. She came up with a brilliant idea: a program allowing municipal utilities to receive forgivable loans. Together, these programs could function like a mini-CEPP, without the penalties, to help utilities start to make progress. I told her I would include them in my conversations with the Hill.

After hanging up, I went back to that first report with Ricketts. We'd laid out six ways to approximate a clean electricity standard through budget reconciliation. One idea caught my eye: federal dollars for states that chose to move faster on cleaning up the grid. If utilities didn't want funding to do the right thing, maybe states would. We'd been talking with Energy Department staffers about the idea, but it didn't have much traction in Congress. Maybe now it would. I wrote a short memo, then called Ricketts to pitch him on it. He was excited. After six years working with Governor Inslee, including leading climate work for the Democratic Governors Association, Ricketts knew that the states were the best place to get shit done. We decided to ask for a big number: $50 billion. And to avoid Manchin killing it again, we aimed for the policy to go through the Environment Committee, rather than the Energy Committee. With $50 billion, I optimistically guessed we might be able to fill half of the hole. I put this idea at the top of my list.

Once I had a rough cut, I shared the draft list with Senator Smith's office and the Congressional Progressive Caucus. Keane Bhatt, who staffed the House caucus, called to talk it through. An alum of Senator Sanders's office, he had been drawn into progressive politics in 2003, radicalized by the US military invasion of Iraq. We'd become friends over the past year. I told him I felt embarrassed by losing the CEPP—like I was a failure. Keane told me he knew what I meant. In April 2019, he helped champion a bipartisan resolution to end the American military's involvement in the Saudi-led war in Yemen. It passed, 54 to 46, in Mitch McConnell's Senate. And then Trump vetoed it. One man had stood in the way of progress. Now 20 million Yemenis' lives were at risk, with many starving. That was the weight he lived with, because he had tried and come up short. Even though he had convinced everyone he needed to in Congress. Because one man with a lot of power said no. Keane's story left me feeling less alone.

Then Keane skimmed the list. "Could you expand this into a memo, with more details about the policy?" he asked. "Oh, and could you get it to me within the hour?" That way it could go on the agenda for Monday's Progressive Caucus meeting. I hung up and got back to work, adding each program's cost and pollution impacts. When I finished the table, the list of programs looked like they might fill the CEPP hole, with nearly the same price tag. I emailed it to Keane, feeling hopeful.

Eight hours later, I heard that the entire Progressive Caucus in the House had my analysis and was using it to try to secure more climate funding. Senator Smith would brief them the next day based on my memo. It was another reminder that with policy work, timing is everything.

By Monday, just three days after Manchin killed the CEPP, Ricketts and I had put together an impromptu coalition to push for these state grants. Folks from across the climate movement were eager to support any efforts to push progress. The group included progressive organizations, like the Sunrise Movement, the Natural Resources Defense Council, Earthjustice, the Sierra Club, and the Working Families Party. But we also had moderate groups, including Third Way, Clean Air Task Force, and Resources for the Future. Energy Innovation joined, too, to answer technical questions. It was an impressive bunch. Together, we decided to request a meeting with Gina at the White House Climate Policy Office to ask for that $50 billion.

Within a few days, we were all assembled, remotely, for the call. From the White House, Gina was there, as was Sonia. Person by person, our crew spoke eloquently to the need for progress in this fragile moment. Varshini Prakash talked about the young people who were angry, saying we needed to support "advocates on the front lines fighting gas and coal plants." Maurice Mitchell, head of the Working Families Party, reminded everyone that we couldn't just let Manchin unilaterally kill the package's central policy. "We need to find new ways to reduce pollution, especially from power plants," he said. Michael Brune, executive director of the Sierra Club, added that "the community wants to fight. They are demanding more. There has to be something that the climate community can celebrate."

Sonia and her White House colleagues listened intently, taking notes. Then Gina spoke. "We are looking for ways to make up for the lost pollution cuts. This seems like one promising idea," she said. "Although, we probably won't get $50 billion." As she pointed out, the Environment Committee had a comparatively small allocation in the budget resolution. She advised us to keep trying to push for this idea through the Energy Committee, which had much more funding. "That's true," Ricketts responded. "But Manchin might just reject this idea too." It was a constructive conversation, and as the call ended, we felt energized. Maybe we could make up for the loss after all.

Meanwhile, Adrian and Tim were also trying to fill the hole CEPP had left behind. For the last two months, they had assumed that Manchin would kill it eventually. So they were already deep in the weeds and had brought Sonia into the effort. Together, they became ton hunters, trying to find policies that either cut lots of pollution for little money or were politically compelling. Tim's model showed that adding more funding for state and local governments to update their model building codes was one solution. He figured that if they put $500 million toward the effort, it might net them 10 MMT: it was nowhere near enough to fill the hole, but it provided the biggest bang for their buck in the whole package.

After talking to a bunch of advocates, Tim came up with a bigger idea. Compared to electricity and transportation, the country was way behind on cleaning up pollution from industry. To make progress, several things needed to happen simultaneously. First, manufacturers needed to adopt new technologies to produce things like clean steel. The bill could include more funding for pilot programs. But if they expected firms to change their behavior, companies needed to know there was a market for these more expensive goods. Tim realized that the federal government could create that market. Once he brought this idea to Adrian and Sonia, they started looking for departments with big procurement budgets. The Department of Transportation, General Services Administration, and Federal Emergency Management Agency were all prime candidates. They proposed adding funding for all these agencies. On top of this, the EPA could use a certification program to clearly label clean products, helping them find other markets. The result of these linked programs was greater than any policy alone. In the ambition tracker, Tim called this strategy "the industrial policy module." He estimated it would fill half of the hole.

Sonia also talked with her boss Gina about our state grants idea. They wanted to make it work. But it soon became clear that we were right: Manchin's Energy Committee would never play ball. Ricketts called the Environment Committee staff, who said they might be able to set aside $5 billion for an EPA program. It was one-tenth of what we proposed—and the wrong agency. But it was something; bit by bit, we were making progress.

Looking over their work, Sonia, Adrian, and Tim found that they had managed to make up for almost all the lost tons, with the one caveat that the uncertainty around their estimates was much bigger. Sonia's one-year-old son had been listening in on their long calls, and noticed his mother was always talking about this strange, abstract idea. He began to parrot her, walking around the house repeating "uncertainty is when you do not know some-ting."

On Sunday, October 24, Ron Klain, the White House chief of staff, drove up to his boss's house outside Wilmington, Delaware. Over his years working for Biden, he'd visited many times. Biden had built the large, colonial-style home in the late 1990s, on the banks of a pond. It was a beautiful morning, the autumn leaves showing their colors, as the president walked Senator Manchin around, giving him the full tour. Ron thought they were showing Manchin a great deal of respect: Biden hadn't had any other members of Congress up there since becoming president. The honor was lost on Manchin. He hadn't even wanted to come, but Ron insisted over the phone that he show up.

Ron understood that if they didn't have a deal with Manchin on reconciliation, the Congressional Progressive Caucus in the House would tank the infrastructure bill. With most of their 95 members voting no, it would fail. Jayapal had made that clear on several occasions. The president wanted to get a gentleman's agreement out of Manchin. If they had that, he could call up Jayapal to assure her they would deliver the climate and care agenda through reconciliation.

Once Senator Schumer arrived, breakfast was served in the library. It was an intimate space, with dark wood-paneled walls, that Biden used as his home office. Brian Deese wandered over to the breakfast buffet, glanced in passing at the bagels, and poured himself a cup of coffee. At meetings like these, all Brian did was drink coffee. He took a seat on one of the plush chocolate leather couches arranged in a horseshoe. And then the discussion began. Together, Ron and Brian aimed to whittle down the Budget Committee's $3.5 trillion agreement into a deal Manchin would support. They went, program by program, through the entire package. It took hours.

The president pressed Manchin to support some modified version of CEPP, but Brian knew the senator wouldn't go for it. Accepting this was difficult for the president. Since the election, it had been his big climate idea. But Manchin was clear that he didn't want to pay utilities for what he claimed they were already doing. And that was the end of it. Brian felt like they'd just lost their anchor on climate and were adrift.

And that's when the yelling began. Schumer was not willing to sacrifice funding for his top priority: affordable housing. There was supposed to be $80 billion for it. Looking directly at Manchin, Schumer said, "We're not doing this bill without housing. That's my number-one issue in New York!"

But Manchin wouldn't hear it. "We're not doing a bunch of checks for people to sit at home and collect free rent," he yelled back. Brian and Ron were used to tense political conversations. But this was unusual. It was like they were brothers screaming at each other in their childhood home, rather than senators sitting in the president's living room.

This went on for almost four hours. And then it was lunchtime, but no lunch was served. Everyone was tired and hungry. Ron and Brian focused on finding what little consensus had emerged between the shouting. On a piece of paper, they wrote down the contours of a deal that it seemed like Manchin might agree to, and passed it to the president.

"I need this, Joe," the president pleaded. "I've got to have this."

Manchin, looking down at the paper, said he could do it. He would help the president get this done. Then they shook hands. It was a deal.

After everyone else left to enjoy what little remained of their weekends, Ron stayed behind with the president to celebrate. They had what they needed to get Jayapal and the rest of the House progressives on board with the infrastructure deal. It was another one of those rare good days.

What Ron failed to realize is that Manchin had a very different understanding of what had just happened. Manchin didn't think they'd made a deal at all.

22

FALLOUT

Late on Friday, October 15, when news broke that CEPP was dead, Nikayla Jefferson was in her apartment in Chicago. She clicked on Coral Davenport's story, read my quote, and felt her heart fall into her stomach. Nikayla was in her mid-twenties and had graduated from UC Santa Barbara, where she'd taken my energy policy class. She'd returned for graduate school but was on a hiatus to organize with the Sunrise Movement to help pass a climate bill. Reading the article that night, she understood immediately how devastating this was.

She sent the story to one of her closest friends at Sunrise, Kidus Girma. She told him they needed to do something to explain to the world what had happened. Something dramatic to capture people's attention. For the past year, they'd tried everything they could think of to increase coverage of the climate bill: organizing rallies, heckling politicians, marching 400 miles. Once again, they needed to raise the stakes. But they both knew there wasn't enough time to organize a large number of participants. Nikayla pulled up the website Beautiful Trouble, with its list of nonviolent protest tactics. She was looking for something big, something that showed sacrifice, something that a small group could do. "What if we did a hunger strike outside the White House?" she said. Kidus thought it was a good idea. They spent the rest of the night calling others at Sunrise, finding a handful of people willing to participate.

Five days later, they stood outside the White House fence. Kidus and four other Sunrise activists were starting their hunger strike. They'd brought five red beach chairs and set them up in a line. One by one, the strikers sat down. On black-and-yellow cardboard, they wrote down their reasons for participating. Kidus's sign read "Hunger striking for my future." As the days passed, the beach chairs were replaced by wheelchairs, the strikers too tired to walk. As lead organizer for the protest, Nikayla's job was ensuring their message was broadcast as far as possible and that they had the medical attention they needed. Shuttling the activists to the hospital, she wondered what she'd gotten herself into. It was brutal watching her friends starve.

I followed the protest with a growing sense of horror as I tried to fill the hole CEPP had left behind. Multiple times a day, my mind wandered to these young people, and I felt a deep sadness. I couldn't shake the thought that I was somehow responsible for their suffering. That I'd given them false hope by trying to pass the clean electricity program. And I couldn't see how and when they would stop the strike. I called Varshini to get her thoughts. She had just visited the group and told me that watching the protest in person was gutting. But she didn't think there was much anyone could do to stop them. They were determined to continue.

I called Nikayla again and again, trying to reach her, hoping there was something I could say to help these kids, but she didn't answer. Her friends were putting their lives on the line, and she was busy making sure they didn't die.

The outcry of sadness and anger these young people embodied helped Adrian, Sonia, Tim, Ricketts, and me add more investments to the package. After the conflict within the environmental movement over the clean electricity program, it was surprising to see the support our idea ultimately had. The criticisms were less surprising. I would scroll Twitter and stumble on hot takes explaining our failure. The program was a utility giveaway. It was too complicated. We should have pushed for a carbon price. When Ricketts and I talked about the range of reactions, we felt frustrated. These critics usually did nothing to help.

Still, Ricketts—of all people—told me to stay positive. He said it was time to sell the climate bill to the movement and the media. He believed strongly that failure would be demoralizing. It would alienate people from the fight. And we needed people, like these Sunrise activists, to keep pushing. "We have to say that this bill, even without the CEPP, is worth fighting for," Ricketts said. "We have to get it over the finish line."

I understood his strategy but wanted to be truthful. And I didn't know if the latest bill would cut enough pollution. Tim and Sonia were still running their numbers. When Sonia and I talked through their findings, she said that with the new additions, the bill could still deliver most of the progress we needed. That conversation left me ready to sell it.

Ricketts and I started hearing there might be a deal any day. Now that Senator Manchin had killed an elephant, he should be ready to pass the bill. At least that was the theory.

As the hunger strike neared the two-week mark, the five activists ended the action on November 2. Their health was deteriorating, and they thought a more direct, confrontational strategy might prove more effective. To keep up their momentum, two days after the strike concluded, Kidus joined a group of activists and went searching for Manchin. It was November 4, and Pelosi had scheduled the House vote on the infrastructure bill for the following day. After that, the climate bill would no longer be moving in parallel. They were running out of time.

Kidus's body felt full of energy—not like he'd just been starving. As Manchin left his houseboat, where he lived while working in DC, the activists followed him. After a few blocks, the senator entered a parking garage, and his security team blocked the door behind him. Kidus realized Manchin was going to drive out. They could keep following him if they found the vehicle ramp and walked down. He told the person next to him to go left while he went right. Kidus found the ramp first and entered the garage, where he saw Manchin behind the wheel of a Maserati. With endorphins surging, he walked over to the driver's side and began talking to the senator. "Is money that important, Joe Manchin? Is it more important than my life, Joe Manchin? Is it more important than your kids' lives?" A dozen other activists surrounded the car. His security tried to push them off so he could drive away. Video of the event went viral. In it, Manchin looks disturbed, perhaps even scared.

The next day, the House passed the infrastructure deal. We'd tried to stall it, getting centrist groups to call moderate offices, but Speaker Pelosi whipped the votes. When I talked to Keane, he told me it was over. Six progressive members, including Ocasio-Cortez, were the only Democratic holdouts against the bill. Friday, November 5, marked the end of "no climate, no deal."

23

BREAKDOWN

On Friday, November 19, exactly two weeks after the infrastructure law passed, the House finally voted on the reconciliation bill. The Build Back Better Act passed on a party-line vote, 220 to 213, with $555 billion in climate investments. It included many ideas from the House committee on the climate crisis. At the last minute, Speaker Pelosi decided to add several social policies to the package, including paid leave for new parents, which Manchin had said he wouldn't support. From her perspective, the president and Schumer had not delivered the two Senate holdouts. Without a firm agreement, she had no obligation to concede what the House wanted. But this pissed off Senator Manchin. He felt like they were boxing him in, making him the bad guy who killed paid leave—seemingly oblivious to the fact that he was that guy.

Now the Senate had the hot potato. Gerry wanted to jam the bill through. If they put it on the floor and threatened to keep the senators until Christmas Eve, he figured it would pass. It wouldn't be pretty, but they could get it done. He thought Manchin wouldn't vote against the motion to proceed.

To make progress, Gerry began holding regular calls with staff throughout the Senate. He led these meetings like a sportscaster, with a polyester golf shirt, wraparound headset, and smooth voice. He called the political plays as he saw them, giving generous shout-outs to staffers, trying to keep their spirits up as the weeks dragged on.

Meanwhile, Adrian worked to nail down the last details with Manchin's team. In early December, his staff sent a list of demands for the climate investments. Adrian read through them, found most workable, and began to tick problems off the list. Then a few days later, another list came. In a normal negotiation, a single party could only make so many requests. Manchin's asks were endless.

They were particularly stuck on electric vehicles. At the time, EV incentives were broken, perversely punishing the major manufacturers that were first to market. While Congress had come close to fixing it in 2018, then-President Trump nixed the idea. Adrian continued working with Schumer to try to solve the problem, but made little progress. Manchin proved to be another obstacle to EVs. He lamented public investment in charging and said the US was overly reliant on batteries from China—both arguments from the fossil fuel industry's playbook.

Days ticked by. The Senate was scheduled to start its break on Thursday, December 9, but Schumer kept the Senators in town, trying to pass the reconciliation package. Manchin's staff sent over yet more demands to Adrian, and they decided to meet to hash things out. Adrian felt like they were making good progress.

The next week, on Tuesday, December 14, Manchin met once again with President Biden in the Oval Office. This time, the senator came prepared with a list of spending that he could support, including climate investments and even some funding for childcare. But President Biden's team didn't think they could get it done before the holidays, not with the remaining procedural hurdles. The White House told Manchin they would release a statement, telling Congress and the world it was time to close up shop for the holidays. They would finish the bill in the new year.

When Manchin read the draft press release the next day, he had a problem. He didn't want to be mentioned by name. Between the parking garage incident, protesters showing up in kayaks at his houseboat, and hecklers on the street, Manchin was under a lot of pressure. He was particularly spooked when his wife was alone in West Virginia and someone showed up at their house. Failing to recognize his own role in elevating his importance, Manchin wanted less focus on him. His chief of staff sent back the document with proposed changes, asking the White House to either remove Manchin's name or add Senator Sinema's. Ron and other senior leaders didn't exactly like Sinema, but they didn't want to throw her under the bus. She wasn't the one holding up the negotiations. They decided to ignore the request, posting what they believed was an innocuous statement. It said the president and Senator Manchin would have ongoing conversations and work to bridge their differences.

When the press release went live on Thursday, Manchin was furious. He texted Biden's senior advisor Steve Ricchetti, saying he was done negotiating. Ricchetti did not take the threat seriously. He called Manchin the next day, assuring him the White House meant no offense and wanted to continue working with him.

Three days later, and exactly one month after the House passed the reconciliation package, Senator Manchin went on Fox News. It was the Sunday before Christmas. He told the world he was killing the entire bill: the climate investments, the childcare policies, the tax reforms, the prescription drug savings. All of it, gone. Even the right-wing host couldn't quite believe what he was hearing, asking Manchin to clarify. "This is a no on this legislation," he said. It was as if he was playing Santa Claus—delivering everyone a lump of coal just in time for the holidays.

Within minutes, the clip was flying around text chains across the climate movement. As Adrian watched it, he couldn't believe what he was seeing. He'd spent weeks with Manchin's team, working through the final issues. Anything Manchin didn't agree with, Adrian would have gladly thrown overboard to salvage the rest of the climate investments. He could not understand why Manchin would walk away now, after all this effort. It didn't make sense. It couldn't possibly be the end of the negotiations. Manchin would be back. Adrian just knew it.

Gradually, his shock and disbelief gave way to a deep sadness. Adrian's wife, watching his depression grow, felt helpless. There was nothing she could do to fix this. She was exhausted, seven months pregnant, with a three-year-old who woke every night. She was just trying to hold it together.

Like the rest of the White House, Sonia had come to accept that they weren't getting the bill done that year. But Manchin's latest pronouncement was more than a setback. It could be the final nail in the coffin. Before the holidays, Gina gathered the climate team for a pep talk. She told Sonia and her colleagues to take a break. They should enjoy the holidays and come back ready to finish negotiations in the new year. She said not to lose faith: the House had just passed the biggest climate bill in American history. That was worth something. They would get it done in the Senate too. "It won't die like this," Gina insisted. "Not after so much work, not with so many investments in the bill that Senator Manchin wanted. We just need more time."

That Christmas was one of the warmest on record in DC. It felt more like fall than winter. Sonia spent the day outside with her son, throwing rocks in the creek by their suburban rental. She felt guilty. Her husband still didn't have a job. She'd put their lives on hold indefinitely.

My holidays weren't much better. I threw myself into Christmas like I hadn't in decades—though my version was decorating a houseplant with paper ornaments. My husband and I also celebrated becoming US citizens a few days before the holidays, after living in the country for more than a decade.

Despite these bright spots, things at home were rough. Because my children were born so early, I was never able to produce enough breast-milk, despite months of effort. Both of my daughters mostly drank the formula the NICU doctors had told us to use. But as weeks passed, it became clear that one of them couldn't digest the cow's milk. On a doctor's advice, we switched them both to a hypoallergenic formula that smelled like rotten cheese. They vomited up the first bottle we gave them. The doctor told us to keep trying—that they couldn't *really* taste it and would acclimatize. It was terrible advice. Each bottle became a power struggle. It took 90 minutes to force one into each child: an excruciating, demoralizing experience. And they both needed eight bottles a day. Even when we switched back to the old formula, they now refused to drink. When an eight-pound baby stops eating, it's terrifying.

The year ended with the Marshall Fire burning a thousand homes in Colorado. It wasn't even wildfire season, and yet it was the most destructive fire in the state's history. Without a climate law, I knew my daughters would grow up with many more fires like these.

ACT II

VICTORY

24

DUNKELFLAUTE

There's a German word often used in clean energy circles: *dunkelflaute*. It means "the dark doldrums," describing a gloomy fog that hangs around in winter. During these periods, there's not enough wind or sun to make energy. That's what it felt like, in January 2022, after Manchin walked away. We were in the dunkelflaute.

With other activists, we brainstormed far-fetched ideas to get the climate bill back on the agenda. Over the holidays, the viral movie *Don't Look Up* was released. The film is a climate allegory, starring Leonardo DiCaprio as a scientist who is ignored as he warns the world an asteroid is coming. "Let's get DiCaprio to testify about the importance of climate action in front of Congress," I said on one call. We imagined him saying: "I'm not a scientist, but I play one on TV!" Everyone laughed. We got a Senate office to ask him, but he didn't bite. We'd need to come up with something else.

As energy prices rose that winter, I told Ricketts we could use the crisis to try to push for climate action. That was how the biggest energy laws had been passed in the 1970s, and again in 1992, 2005, and 2009. Crisis was the only thing that seemed to compel Congress. Ricketts understood immediately—he'd seen it firsthand with the 2009 rescue plan. And it wasn't a hard story to tell. That January, electricity prices jumped 8 percent year-over-year. And the price of gasoline rose almost 10 percent by the end of the month. We started talking more about how clean energy and electrification could lower Americans' energy bills.

Ricketts's big idea was simple: fake it until you make it. We would just tell every journalist that we were seeing momentum. We pointed out that Manchin had said no to *that* bill, not to *any* bill. Eventually, he might say something ambiguous, and we could spin that into something positive. Ricketts started hearing rumors that other environmental groups were laughing at this strategy, talking about it as rather deranged. But he didn't care. He'd rather look like an idiot than a cynic. He knew that when people stopped trying, that's when we'd fail.

We weren't the only ones still trying. For over a year, Christy Goldfuss had joined a daily call, Monday through Friday, with a handful of her closest colleagues, including Ricketts. It served as a war room for the climate bill, including the heads of several big nonprofits and prominent Democratic fundraisers. Each day, they talked through the press coverage, tactics to respond, and key messages to convey. Christy had served in the Obama Administration, running an environmental office within the White House. She was close with John Podesta, who was chief of staff to President Bill Clinton, a senior counsel to President Obama, and then chaired Hillary Clinton's presidential campaign. Christy and John worked together at the think tank he'd founded, the Center for American Progress, and both cared deeply about passing a climate bill.

During the Trump administration, Christy had been working on a new strategy to ally labor, environmental justice, and climate groups together. Over several years, she'd built a team that included Luke Bassett, who hailed from West Virginia and went on to join Manchin's team. She'd also been supporting efforts to bring environmental justice groups into closer dialogue with the biggest climate groups. And she hired veteran labor organizers to think through policies to get unions on board. Christy's ideas were working—the coalition was bigger this time. She could see progress. That is, until Manchin pulled the plug on the negotiations in December.

Their daily calls took a dark turn in January. Everyone's patience was thin, and the group began sniping at one another. John was skeptical that the climate bill would come back to life. But Christy knew they had to keep trying. She'd started to look ahead to April, wondering if Earth Day was their chance to put attention back on climate. Though that was months away.

At the end of January, Senator Ed Markey from Massachusetts invited us to meet. He began the call with an impromptu virtual tour of his office.

By then, he'd served in Congress for almost 50 years, so it was packed with photographs and memorabilia. On one wall hung a large solar panel from the 1980s. Back when he served in the House, Markey organized a renewable energy fair for companies to show Congress their latest inventions. Chevron brought this solar photovoltaic panel, then threw it in the garbage, where Markey fished it out. As the years passed, he asked his staff to add stickers to the panel, showing the latest stats on renewable energy growth. It was a reminder that progress was happening. And the panel itself proved it: Markey loved to point out that a modern solar panel could be ten times smaller and still generate the same amount of power.

He'd also stuck the Sunrise Movement logo on the panel, as a reminder of just how important the climate movement was to progress. After their sit-in with Representative Ocasio-Cortez at Speaker Pelosi's office in 2018, Markey backed the group's ideas, co-sponsoring the Green New Deal resolution. He'd grown particularly close to Varshini Prakash, who was from just outside of Boston, even taking her as his guest to the State of the Union in 2019. If it wasn't for these young people, he probably wouldn't still be a senator. His reelection in 2020 was tough, with Congressman Joe Kennedy mounting a primary challenge. Markey started from behind, down double-digits in the polls. And then, Sunrise endorsed him. They worked with his campaign to craft a viral ad that showed Markey as a fighter for the people. Sunrise also worked the phones, reaching 250,000 people, and his youth support surged. He won by ten points.

A year and a half later, Markey was one of 50 Democratic senators. And when we met with him that day he had a pep talk ready. Markey said we were some of the last people who still believed the climate bill could pass. But his years of experience had taught him that this was the most treacherous time for legislation. Losing momentum was deadly. "Anything could knock the bill off course now," Markey said. We all thought back to the Waxman-Markey bill, which bore his name. It was like a ghost hovering over the conversation. Here we were again: a climate bill had passed the House and might never make it through the Senate.

Other senators might find it frustrating that a failed bill haunted them for a decade. But Markey didn't mind. He wanted to figure out what went wrong and not repeat the same mistake. "With legislation, there's no guarantee except one," Markey said. "If you don't build the movement, you will not win." He told us to keep working with our networks of advocates. He said that hope only existed when people took action. We needed to keep going. We had to keep telling folks that victory was still possible. Because that was the only way we might still win.

As a new mother, I was struggling. Feeding one of my daughters continued to be a ninety-minute battle. That month, when my parents visited to help us out, my father said he thought the girls were conditioned to avoid their bottles. When they saw one, they turned their heads away. A lightbulb went off: I did some research and quickly discovered my daughters had developed bottle aversion—a common problem for NICU babies who have eaten from feeding tubes for long periods or for infants forced to eat disgusting formula. No doctor had ever mentioned this.

After binge-reading a book on the topic, we slowly reconditioned them to drink milk again. The basic idea was: let the baby decide whether to eat and never force. I also switched them to goat's milk formula, thinking it might be more digestible. It was a painful process that took weeks, but they recovered well, began eating, and gained weight again.

At the end of January, our daughters turned six months old. They were still tiny, weighing less than 12 pounds, the size of a large newborn. After a hard few months, I decided we would celebrate their half birthday and bought a cake. That day, I also resolved to quit pumping, after six months of effort.

With babies, time moves differently. As a mother, I began to slow down to their pace. Watching these tiny beings made me notice the smallest changes. Look, now her eyes can track a toy above her head. Look, her movements have grown less jerky. Look, her eyes are open, more aware. I tuned into their slower, sleepier wavelength.

I also became more aware of hazards. We shoved plastic into every power outlet and worried about sharp edges on coffee tables. It wasn't even possible to leave a hospital without a car seat manufactured within the last decade. But it also struck me that these deeply ingrained babyproofing practices, which protect newborns in their vulnerable early years, were narrow in their approach. Parents imagined what they would do if their house was on fire: how they would grab the kids first, then their pets, then the photo albums if they had time. But few people acted like their kids' house was already on fire: the planet they called home. They didn't act like that was an emergency.

On February 1, Manchin told reporters that Build Back Better, the bill that passed the House in November with social and climate investments, was "dead." I thought back to his old campaign ad, where he shot the Waxman-Markey bill. It wasn't a metaphor: he aimed to kill these climate bills. Manchin said he had not talked to the White House since December and any bill would need to start from scratch. A week later, he went on TV with Senator Lisa Murkowski, a close Republican colleague from Alaska. He reaffirmed that the bill was dead, and said if any new package was proposed it had to go through committee.

We knew what it was like to lose parts of the bill, but the whole thing? It was an outcome I hadn't truly grappled with. It felt like staring into an abyss.

25

REBIRTH

There are a thousand ways to fail. Much easier to fall off a cliff than scale a mountain. A senator dies, the Democrats lose their supermajority, and there goes the climate bill. A candidate has an affair, loses his election, and there goes the Democrats' 51st Senate seat. A press release is posted, a senator is angered, and there goes the president's agenda. There are a thousand ways to fail. And perhaps just one narrow path to success.

For Adrian, failure was constantly looming. The longer he worked in Congress, the more anxious he became. And for good reason. In early 2021, when Democrats prioritized the COVID bill and punted on climate, Adrian thought, "We're screwed." But he forced himself to believe that a pandemic was just an unusual situation. There would be a second reconciliation bill.

Then, in November, when the bipartisan infrastructure deal pulled ahead of the climate package, Adrian thought, "We're screwed." But he convinced himself this was always the plan. Everything bipartisan would move first, and then they would turn to the climate investments. "This is inevitable," he told himself. "This is how we're going to get all of our caucus to be happy."

Again, in late December, when Manchin blew up the talks on Fox News, Adrian thought, "We're screwed." But he refused to dwell on the feeling. Instead, he picked apart Manchin's comments, telling himself he was just declining *that* package. He would eventually come around to something else. He had to.

Maybe Adrian was being delusional, creating lies to preserve his sanity. But it worked: he kept pushing to pass the climate bill. "There were moments where things were going in a very bad direction, but we had the sense that failure was not an option," Adrian told me years later. "Even though failure was very much an option—we were always imagining a million ways that we could fail. But it was really hard to accept that we would do nothing on climate."

One reason Adrian remained optimistic was that he had a friend on Manchin's team: Luke Bassett. They'd gone to Yale together, studying climate policy. After graduation, they both moved to DC and became closer. When Adrian joined Schumer's team in 2019, Luke mentioned he was considering a job with Manchin on the Energy Committee. Hailing from West Virginia, Luke's family had known Manchin's for a long time—it wasn't a huge state. Adrian encouraged him to take the job, knowing that having him there would prove enormously helpful if Democrats won a trifecta in 2020. When Luke joined the committee, Adrian was thrilled, and they'd collaborated closely for the last two years.

That January, Adrian's wife was eight months pregnant with their second child. When she spoke to her parents at night, Adrian would inevitably be on a work call. Her parents couldn't understand why: it was past 11:00 p.m., how could he still be working? But that's what the job demanded. She told them to stop asking about it.

For the past three years, Adrian had regularly put in eighty-hour weeks. Like a firefighter or doctor, he was usually on the clock. Senator Schumer was known for calling his staff daily—calls lasting anywhere from 30 seconds to several hours. When other offices used recesses to take a break, Adrian just kept going. Recess didn't exist for Schumer's team. With Democrats in control, every day counted.

Adrian lived 30 minutes outside the city, in a beautiful home adorned with solar panels, but was often away at the office for long hours. He claimed to hate wearing suits, especially in the humid DC summers. But this was only partially true. In a suit, he looked sharp and polished, like a man with important work to do. As the years passed, his wife got used to him getting dressed in work clothes on Saturdays. She didn't know where the man she married had gone. That guy loved to laugh and wear sweatpants. This guy was always stressed and wearing a tie.

Most nights, when Adrian got home, he would try to take a two-hour break, to cook dinner and spend time with his daughter. But it wasn't always possible. His job was relentless. It wore on him mentally and strained his marriage. His wife wanted him to quit.

There were days when Adrian wanted to quit too. But the way he saw it, his work protected his family. He gave up time at home to ensure their future. "To be a part of these historic pieces of legislation, you put yourself on the line, and you make sacrifices because it's important," Adrian told me. "It's much bigger than you, and your needs, or your sleep. That shit doesn't matter compared to the scope of what you're working on. And everybody has that attitude, or they don't last very long."

Adrian believed the most important thing he could do for his children was keep working to protect their future. If he couldn't get a climate package passed, he'd have failed humanity. But even more acutely, he would know he'd failed his kids. That heavy sense of personal responsibility ticked like a metronome in his head, pushing him. Even when it seemed like his work might lead nowhere.

Their son was born in mid-February, almost a month early. They agreed Adrian would take his paternity leave after his wife's leave finished. He told her the bill should pass by June at the very latest. As soon as it was done, he'd take a break. She had three months, so there wouldn't be much of a gap. In the meantime, he kept up his grueling schedule.

The rest of the Schumer team was in a similar place, working long hours. Tim continued to meet with Sonia, calibrating the ambition tracker. While their work had begun as geeky fun, it now felt like studying for the SATs. They had to learn every single detail of the package so they could recall it from memory when needed. They knew they'd face pressure to cut some programs, and they wanted to be prepared to explain why specific investments needed to stay. They both still believed there would be a bill, eventually. Even if it made them look naive.

Gerry felt like he was in the film *Groundhog Day*. Every morning, he would wake up, take a shower, and stand under the water thinking: "Is this ever going to end?" He was dogged by feelings of regret and began blaming himself for Manchin going AWOL. They shouldn't have let Sanders start the negotiation at $6 trillion. Or even gone with the $3.5 trillion from the Budget Committee. They'd messed the whole thing up.

His mind kept circling back to a meeting in the Oval Office that fall. Speaking out of turn, Gerry had said: "Mr. President, we should consider just doing $1.5 trillion right now. The clock is ticking. Someone could die and we could lose the majority. Let's just do it." He'd run the idea by Brian Deese ahead of time, and they'd agreed it would be helpful for Gerry to propose this. But Speaker Pelosi, also in the room, did not like this plan. The House wasn't ready to cut the bill that drastically. They needed to stick with the process. Gerry couldn't help but feel he should have pushed harder. Now the bill might never pass, and he would struggle to forgive himself if that happened.

Senator Schumer spoke with Ron Klain, Biden's chief of staff, in January. Ron emphasized that the president's top priority was the climate investments. They agreed that Schumer should take the lead on the negotiation going forward, because, quite frankly, Ron didn't trust Manchin anymore. "This needs to be wrapped up before July 4," Ron told Schumer. If he failed to meet that deadline, the White House would move on, turning to executive action to make progress on climate. Schumer assured him he'd get it done.

When Schumer spoke with him again, Manchin was still angry at the White House. But Manchin also told Schumer that he was open to revisiting parts of the package during the "second quarter." Schumer fixated on April: that was when Manchin would be ready to negotiate again. In the meantime, Schumer decided to send Senator Tom Carper, chair of the Environment Committee, to restart negotiations. He had a good relationship with Manchin—they were both born in West Virginia and shared a pro-business sensibility. They focused their discussions on the methane fee—a plan to charge oil and natural gas facilities for releasing methane beyond certain thresholds, incentivizing them to reduce their pollution over time. This was one of the outstanding thorny issues. It felt like Carper was making progress.

Even a step removed from the negotiations, the White House was still trying to convince Manchin to pass the bill. Business support for the package seemed key. As part of this strategy, the meeting with utility executives, which Sonia had been trying to organize for months, finally got on the president's calendar—albeit four months too late to save the CEPP. Still,

Sonia knew that photographs of the president sitting down with utility leaders might generate momentum.

The meeting took place on February 9, with ten utility leaders, including American Electric Power, which supplied electricity to West Virginia. Biden began by thanking the executives. He said the electricity they provided to Americans was the backbone of our current and future economy. Perhaps surprised to be greeted with such warmth, the CEOs leaned in as he spoke. They were being lauded by the President of the United States of America. That didn't happen every day.

Listening to the conversation, Sonia found the president impressive. Maybe it came from his decades in the Senate, but he tended to thrive in face-to-face negotiations. The man people saw in big speeches wasn't the same man she saw in these smaller rooms, at least not back then.

In the briefing materials, Sonia asked the president to focus on getting the grid to 80 percent clean power by 2030. She worried that the CEOs would recoil if he started talking about 100 percent clean power. They were constantly fixated on how hard that last 20 percent was. Biden got quickly into these details, saying this plan would create good-paying jobs in the United States while cutting pollution. And he told the utility companies they simply had to treat climate change as the serious threat it was. Even when he went off script, Sonia thought he was making an important point.

The meeting lasted much longer than scheduled, running well over an hour. As she walked back to her office, Sonia's mind began to loop. It had been a productive meeting, but the utilities had not committed to anything concrete. "We'll get it next time," she thought. "We're still going to get it. We're just going to get it another way. We're going to find another way to get it because we've got to get it. We're just going to get it."

Two weeks later, on February 24, Russian President Vladimir Putin invaded Ukraine. Energy prices shot through the roof. The following week, Senator Manchin told the press he was open to passing an energy and climate bill.

26

THE DEFENSE PRODUCTION ACT

When Putin invaded Ukraine, Sonia was on the phone well past midnight, talking to her White House colleagues on the National Security Council. They were receiving calls from European leaders planning to wean themselves off Russian gas as fast as possible. They insisted they needed America to supply more natural gas. Sonia didn't want this moment to lock in more fossil fuel infrastructure. She started looking for ways to support the Europeans without rubber-stamping new liquefied natural gas terminals. Besides, those projects wouldn't help in the short run, since they would take years to come online.

Sonia wanted her White House colleagues to think about all the options on the table. To make each alternative comparable, she needed to know how much energy it could supply and how long it would take. It was a mad scramble. She dug up data on how much gas Europe needed without Russian supplies, current American gas contracts, and the status of existing LNG terminals.

On February 27, just three days after the invasion, climate activist Bill McKibben called on the federal government to invoke the Defense Production Act of 1950. Created during the Korean War, the law allows the president to intervene in the economy in times of emergency. Bill was saying that Biden could invoke it to scale up domestic manufacturing of heat pumps to ship to Europe. I thought this was brilliant. Heat pumps would let Europeans break their dependence on Russian gas, and cut pollution along the way.

To generate buzz around the idea, the team at Rewiring America put out retro-style wartime posters with slogans like "Send Putin away, get a heat pump today" and "Heed the cry, electrify!" Meanwhile, we started digging into the data to understand how fast manufacturing could be scaled up. One of the biggest challenges was that Europeans primarily used hydronic heat pumps, which warmed water to heat buildings, whereas Americans used air-source heat pumps, which directly warmed the air. There wasn't any existing manufacturing capacity in the US for hydronic heat pumps. But boiler manufacturers could retool and begin to produce this technology. These ideas began to catch fire: when the International Energy Agency published a ten-point plan to reduce EU dependence on Russian gas, heat pumps made their list.

A few days later, on March 9, Rewiring America released the "Electrify for Peace" plan. While our ideas were received well, with widespread media coverage, some feedback from the climate movement was strange. Folks made bizarre comments, like "the US isn't a democracy, so how can you call for heat pumps to protect democracy," and "this conflict isn't about fossil fuels, so you're exploiting a tragedy." It foreshadowed things to come.

Whether or not the US could actually manufacture and ship heat pumps to Europe fast enough, like Sonia, we wanted policymakers to realize there were multiple ways to deal with the gas shortage in Europe. In the long run, the most important way to support European energy independence was with clean energy and electrification. And that would only happen if we passed the climate investments.

I also believed that media attention would help us secure support for American heat pump manufacturing. For over a year, we had been brainstorming policy approaches to make that happen with very little traction. This energy crisis provided an opportunity to change the narrative. If the United States wanted to be energy independent, we needed to make clean technologies here. But when Congress approved an emergency package of funding for Ukraine, none of it went toward heat pumps. We'd have to keep pushing to get it through the reconciliation bill—assuming that ever passed.

Suddenly, political space for the climate bill began to open. As Ricketts and I had predicted, an energy crisis provided opportunity. The week after Putin invaded Ukraine, oil prices jumped another 18 percent. If this trend continued, inflation would spike, since fossil fuels are used to produce many goods and services. It was then that Manchin told the media he was open to restarting work on an energy and climate bill.

When I read about Manchin's renewed interest, I was ecstatic. The next day, I called Ricketts to talk strategy. We were both surprised more climate groups hadn't seized on this moment to push for the bill. We decided it might be a good time to renew pressure on Schumer and, with others at Evergreen, brainstormed a protest in front of his home in Brooklyn. When we reached out to folks at Sunrise's New York City hub, they were eager to make it happen.

On Monday, March 14, Sunrise and other local climate groups protested at Schumer's house. They were joined by actress Ilana Glazer, who livestreamed the event on Instagram. Although the event received little media coverage, that was beside the point. The aim was to get Schumer's attention. Manchin had a tendency to change his mind. If he was open to an energy package now, Schumer should move as fast as possible. That same day, a coalition of 120 groups, including the Working Families Party, released a letter calling on Schumer to act.

At the end of that week, on Friday, March 18, Senator Joe Manchin traveled with Energy Secretary Jennifer Granholm and Interior Secretary Deb Haaland to West Virginia for a full day of events. The White House sent Brian Deese, who had built a good relationship with Manchin over the past year. When someone in the administration annoyed the senator, he'd taken to calling Brian to vent. "What you guys just did was crazy!" Manchin would yell over the phone. "I don't know, Brian. Maybe this whole thing is just impossible. It feels like I'm not aligned with the White House." Brian, an even-keeled guy, would listen and try to bring the temperature down. "Senator, you've been in politics a lot longer than I have," he told Manchin. "But sometimes people just say things. It doesn't mean that's what we're doing." Brian seemed to reassure him, and he would usually hang up in a better mood. Brian became the unofficial White House liaison to Manchin.

This was Secretary Granholm's second trip with Manchin to the Mountain State. They'd first traveled to West Virginia ten months earlier, in June 2021, for a press conference at a steel plant working on offshore wind. The goal was to highlight how a coal state could help lead the clean energy transition. They had so much fun that they ended the evening at a karaoke bar in Morgantown, drinking bourbon and singing "Country

Roads." That's when Granholm learned Manchin's favorite song at the time was "S.O.B." by Nathaniel Rateliff & the Night Sweats. The senator stood on stage, passionately belting out the blues chorus—a ballad about a man trying to get sober. Granholm and both of their staff couldn't help but laugh. Manchin was a lot of fun.

This second trip, in March 2022, had a packed agenda. First, Manchin, Granholm, Haaland, and Brian had breakfast with Governor Jim Justice at the Governor's Mansion—his bulldog under the table continuously popping his head into Brian's lap. Then they traveled to the Robert C. Byrd Institute at Marshall University to announce that a battery company would build a new factory and retrain displaced mine workers, in partnership with the union. At the press conference, Granholm added that West Virginia would be "well positioned" to secure federal funding from the Biden administration's new retraining programs.

Their last stop was a roundtable with businesses eager to build more clean energy projects in-state. Granholm's staff spent months preparing for this meeting, believing it might convince Manchin to support the stalled climate package. As they headed to their cars to drive the two minutes to this third meeting, Manchin and Granholm were talking and smiling. But when she joined her staff and shut the car door behind her, Granholm turned serious. "Well," she said, "Senator Manchin isn't coming to the roundtable." Her staff was confused, then furious. They'd spent so much time preparing every participant, ensuring each would say the right thing and use their time wisely. Even Manchin's wife was going to attend. Granholm's staff immediately called up Manchin's lead on the ground and explained the situation. "He says he's not coming. He has to be there. We have done everything he's asked." His staff was equally surprised. "I'll call you back," they said, before hanging up. About two minutes later, they told Granholm's staff: "He'll be there."

For several hours, Manchin, Granholm, Haaland, and Brian listened to businesses explain how they were going to build clean energy jobs in West Virginia. Initially, the companies just introduced themselves, explaining what they were working on. But Manchin cut them off. "Listen, guys," he said. "You've got two cabinet secretaries here. You've got a high-ranking member of the White House. And you have me. You have all of our time—so use it. I don't want to hear about your companies. I want to hear what you need from me." This turned the whole meeting around. For the rest of the afternoon, they got into specifics: like how a manufacturing tax credit could help them build new factories and federal grants could spur retraining.

When one company said they wanted to have tax credits available even if they didn't owe the federal government money, Manchin interjected. "But wait. If you're a nonprofit or a co-op, I get it. You don't owe the government taxes. But if you're a company, you're profitable. You have tax liability. Why should the government be writing you a check?" Although he didn't agree with the senator, Brian leaned in to listen. He could see Manchin had specific, practical concerns. Brian could work with that.

When it was over, Manchin seemed moved. These companies were talking about big investments that would help his state become a real player in the clean energy economy. And that didn't just mean solar: it meant critical minerals and repurposing coal infrastructure. Manchin had heard a lot of bullshit in his time, but this seemed real.

The rest of the day, Granholm and Manchin were headed in opposite directions. She was scheduled to tour a community solar project across town, while he was heading east with Secretary Haaland and Brian for more tours. Manchin asked Granholm to come along, but she knew she couldn't—cabinet secretaries never controlled their own schedules.

As they drove to an abandoned mine being reclaimed with funding from the infrastructure law, Brian talked to Manchin about his comments at the roundtable. He told the senator he understood his concerns with using "direct pay" to give private companies federal checks. "But, you know, sometimes companies don't owe enough in taxes to use the credits, and they need to finance it, which eats up a lot of the value," Brian said. Manchin listened. "Well, maybe there's another way we can solve that problem," he replied.

Their day ended just outside New River Gorge National Park, which is when Manchin asked Brian: "Do you want to go ziplining?" "Sure!" Brian told him, not wanting to pass up a bonding opportunity. After posing for photos in their orange helmets, they each jumped off the platform, and went flying through the air. That night, Brian stayed up late, sitting around a bonfire with the senator and his staff, just talking. He knew that was the entire point of this trip: he had to keep listening to Manchin.

On April 1, a colleague at Rewiring America asked for a last-minute meeting to share some news. "This just stays between us. But I heard from one of our lobbyists that a European recently showed Manchin a heat pump. And he liked it! He thinks it's innovative technology!"

"This better not be an April Fools' joke," I replied.

"It isn't," she exclaimed. Senator Manchin had taken yet another trip with Secretary Granholm, this time to Europe in the wake of the war in Ukraine. The continent was buzzing with ideas to reduce its reliance on Russian gas. And heat pumps were a popular solution. On that trip, someone had the genius idea to show Manchin a heat pump, and he'd reacted positively. It was a testament to the power of giving policymakers tangible, firsthand experience with a technology.

At the White House, Sonia tried to get the president to invoke the Defense Production Act to ramp up domestic manufacturing of clean technologies, including heat pumps. At the end of the month, Biden did use the law to boost critical minerals for batteries—notably on the same day Manchin was holding a committee hearing on the topic. But Sonia wanted the president to use it more broadly for clean energy. Outside of the climate office, she was not exactly finding a warm reception for the idea. The national security folks told her it could take 18 months just to write up the justification. To make it happen, she would need to work with the Department of Defense, and there was just one guy assigned to the file—and he was busy with "more important" things. But Sonia kept working. The fact was, energy was a national security issue.

At least she was undermining the National Security Council's first instinct that the solution to Europe's crisis was building a bunch of LNG terminals. Although the US-EU agreement, released on March 25, focused extensively on new gas supplies, it also named heat pumps, renewable energy, and energy efficiency as other solutions. It wasn't a full victory, but Sonia knew she had pushed things in a better direction.

There wasn't exactly time to celebrate. At the White House, there were always new problems. Now Sonia was worried that, in a few months, the United States might not have enough solar panels.

27

SOLAR INTERFERENCE

After a year in DC, Sonia was tired. She still lived in an expensive suburban rental, with almost no furniture, isolated from friends and family, waiting for the climate bill to pass. She worried her son was growing up without a community. Her husband, who stayed home to take care of him, still didn't have a job. As the months passed, they both grew frustrated. Her husband, a talented musician, began composing funny songs about Manchin on his guitar, and belting them out to blow off steam at the senator keeping them in DC. Laughing together was cathartic.

She had hoped to stay only until the climate bill passed, but Sonia didn't know when—or if—that would happen. In any administration, the first six months are critical to moving a president's agenda through Congress. That time was long gone, and little of Biden's vision was now law. Sonia couldn't ask her family to wait forever for this bill to pass. She knew it was time to move.

Sonia asked her boss for a quick call. Gina McCarthy arrived to the meeting with a worried look, like she was preparing for the worst. Sonia didn't know that, at the White House, asking for a fifteen-minute call usually meant someone was quitting. "I just have this question," Sonia began, a bit nervous. "I feel like I've made all these relationships across the federal government, and I now have a really good ability to stay connected with people. So much of the time I'm just sharing an office with a colleague as we talk over each other on Zoom calls. What would you think if I worked remotely instead?" When she finally got to the punchline, Gina's face softened. She wasn't quitting after all. Gina said she would need to think it through, since White House policy now discouraged remote work. A mere ten minutes later, she called back and said it was approved. Sonia and her husband decided they would move to New Mexico in early April. It was where he grew up, and his family was still there.

But before she could start packing, a colleague at the National Security Council called to put something on her radar. On April 1, the Commerce Department began investigating a complaint from a small solar manufacturer in the San Francisco Bay Area. It alleged that cheap solar panels from China were making their way to the US through Southeast Asia, circumventing existing tariffs. It was up to the Commerce Department to investigate, as legally, the White House was not supposed to weigh in. The problem was, a decision could take years. In the interim, any panels brought into the country could accrue retroactive tariffs at an unknown rate. This uncertainty would create chaos for the industry—stopping up to 80 percent of solar imports overnight.

Sonia understood the stakes: this might not just devastate the solar industry but endanger the entire electricity grid. In 2021, over a third of new capacity came from solar. She put together a memo for White House leadership, and they quickly understood the urgency. She was told to figure out a solution.

I heard about the problem a few days after Sonia, when a friend working in the Senate flagged it. Once I started digging into the details, it seemed preposterous. How could the Biden administration shut down solar imports? But when I talked to the solar industry association, they were freaked out. If this didn't get resolved quickly, their members would have to lay off workers. Despite the gravity of the situation, they were struggling to get media attention. And they had no idea how the White House would fix it.

When I reached Sonia on the phone, she was despondent. "It's really bad," she told me. We were already so far behind on getting the climate legislation passed. And now the Commerce Department could block almost all solar panels from entering the country. It was absurd. We agreed to stay in touch, and I got to work drumming up news coverage.

That spring, our networks of advocates started to share more hopeful updates. On our clean energy network call in early April, Sam Krasnow, who worked at the Natural Resources Defense Council, shared that he'd just finished up a day on the Hill. He relayed that everyone seemed to be in positive spirits—despite the literal rabid fox running around the complex that bit half a dozen people that day, including a lawmaker. While he was walking around the Hart Senate Building, with its modernist white walls, he passed Senator Manchin in the hallway. Ever a man of action and optimism, my friend approached the senator while the rest of his group fled for the elevator bank, leaving him alone for a one-on-one conversation with the man of the moment. "Senator Manchin," he called out. "Thank you so much for your work on the clean energy tax credits. They are good for everybody." Senator Manchin replied, "Thank you, buddy. We're working on it." Then he patted Sam's shoulder, smiled, and sauntered off. When Sam met with Senator Tina Smith, she similarly reported that she'd spoken recently with Manchin. He was in a good place and wanted to make a deal happen.

These little glimmers acted as bursts of helium, buoying us all. When Justice Ketanji Brown Jackson was confirmed on April 7, in the middle of one of our coalition calls, we were overjoyed. Not just because the first Black woman had finally been appointed to the Supreme Court. But also because it meant that Congress could focus on other things. The Senate planned to return after a two-week recess on April 25, and finally turn to the reconciliation package. We remained hopeful that the bill would keep most of the House's $555 billion in climate investments intact—and not include too many fossil fuel poison pills. Even Senator Manchin began telling the press he wanted to get the bill done before summer.

Meanwhile, Christy Goldfuss continued joining those daily calls with Ricketts, John Podesta, and other climate leaders. Christy had decided to help organize a big Earth Day rally to try to keep attention on the bill. There was already a large network behind the event, but they needed more people to pull it off. When Sunrise confirmed they would join, Christy knew it would be a valuable event to invest time in.

John thought focusing on this was misguided—and didn't hesitate to share that opinion. He had organized huge Earth Day rallies in the past and knew they were behind schedule if they wanted this one to make a splash. Besides, how would this help convince Manchin? Christy saw this as a fair point, but she also knew they needed something to help morale in the climate movement. At least an event would bring folks together.

Long campaigns required unity. People had to feel like they were working together toward a goal and making progress, whether or not it was true. And she thought it was a more productive use of energy than more protests, which might backfire. There were folks targeting coal plants and proposed pipelines in West Virginia. Manchin didn't tend to respond to pressure in the way these activists wanted.

While I was hesitant to take a break while momentum was growing, I was invited to travel to Hawai'i to meet the governor and give a talk at the University of Hawai'i at Mānoa. We decided to bring our eight-month-old daughters, tacking on a few days at the end, making it our first family vacation. During the trip, I went on the radio, talked with climate advocates, and met with legislators, learning about their work to drive change. And while these meetings on Hawai'i's energy system were fascinating, I was most excited about the snorkeling. To see an iridescent purple coral or a highlighter-bright anemone feels otherworldly. Sadly, visiting a reef is also one of the clearest ways to see climate change in action. Our oceans have been absorbing around one-third of our carbon pollution, masking the full greenhouse effect for decades. But this has come at a steep cost: ocean acidity has risen over time. Corals have paid the price, dying at unprecedented rates. And yet there is still so much beauty left.

Just ten miles east of Honolulu, there's a protected reef called Hanauma Bay. Dense and diverse, it is home to triggerfish, tang, surgeonfish, parrotfish, and even Hawaiian Green sea turtles. That warm, bright morning in April, as I floated among the coral, admiring the magical underwater world, I bumped into a fish. Or, rather, it bumped into me. Seemingly out of nowhere, a medium-sized fish started to swim aggressively toward me. It came at my face, then my feet. Floating in the water, with a fish attacking me, was disorienting and confusing. I swam over to my husband and told him, "I think I upset a fish, I don't know what I did." With curiosity, I returned to the spot, and the same thing happened again. In that moment, understanding that the fish was being aggressive, an unexpected thought came to me: "This fish is trying to tell me something: it's trying to tell me to work harder. I need to do more to stop the climate crisis."

Afterward, I stopped by the interpretive center, and told the woman at the booth, "I had this interesting experience with this fish . . ." I could hardly finish my sentence before she interjected, picking up a well-worn photograph, "Oh yeah, did it look like this?" Clearly, she heard the same story from oblivious tourists a dozen times a day. She said this particular fish is called the humuhumunukunukuāpua'a, or the humuhumu for short. It's beloved in Hawai'i and became the official state fish after schoolchildren ran a successful campaign in the 1980s. It is distinctly marked, with a black band along its torso and a neon sideways V near its tail. She explained that it was spawning season for the humuhumu, when they laid eggs. And after those eggs are made, the fish wants to protect its nest. It feels threatened by unwitting snorkelers potentially stepping on its precious offspring-to-be.

It occurred to me that my thought wasn't so strange after all. The fish was asking me to protect its home. And its home was not just that patch of eggs but the entire coral reef. What that reef truly needed, to continue to be the home for the humuhumu's offspring, was a stable climate. Maybe the fish *was* telling me to work harder to get the damn bill passed.

During her final days in DC, Sonia raced to try to solve the solar problem. It felt like assembling a puzzle without any guide to what the final picture would look like. She needed a solution that was credible and fit within the legal and political constraints. To figure it out, she was working with the White House National Economic Council, Department of Energy, and Department of Justice. Working from inside the White House gave Sonia a panoramic view. She could ask career civil servants to tell her where the levers were hidden and how to pull them. That was the magic of a White House job: you could bring in the best and the brightest across the government to solve a problem.

But it could also be frustrating. As Sonia started to hone in on a series of actions, she needed to get sign-off from several lawyers. And they wanted to go through her proposal, line by line, often spending half an hour discussing a single word. They needed to ensure that they were on sound legal footing, because whatever they did would face scrutiny and likely legal challenges. But there was also a perfectionism that bordered on dysfunctional. Knowing that every day this took to resolve created more panic and chaos in the solar industry made the process maddening.

On our next call, Sonia told me she'd come up with an idea for how to solve the problem. I asked her to tell me what it was, but she wouldn't give any hints. I hate not knowing things. So we began to play a game of 20 questions: I would put together my latest theory, send it to Sonia, and ask, "Is this it?" She would answer in a single word: no. No matter what I guessed, the answer was no. It would be six weeks until I found out her plan.

Sonia made it to New Mexico by April 9, just in time for her father-in-law's seventieth birthday party. The night was warm, and they stayed out late dancing, her light-up shoes illuminating the darkness. Her son was surrounded by his kin. They were happy. But in the back of Sonia's mind, a shadow hovered. She knew she'd left DC without accomplishing most of her to-do list. Even though her old office was 1,700 miles away, she was determined to keep working to check everything off that list.

28

MADE IN AMERICA

On Thursday, April 7, Senator Manchin walked into the Dirksen building's third-floor hearing room and took a seat behind his "Chairman" placard. He was wearing a dark pinstripe suit with a silver tie. On his right lapel, he'd pinned the American and Ukrainian flags, symbols of his opposition to the Russian invasion.

Manchin was hosting a second committee hearing on critical minerals. With his reading glasses balanced on the tip of his nose, he began the opening remarks his staff had prepared. Halfway through, they pulled out a giant printed chart, which the senator pointed to, demonstrating how recycling required far less material than mining. And then he began talking about Superman, of all things—reminding the room how Kryptonite could render the man of steel powerless. Manchin, baffled to be referencing a comic book character in a Senate hearing, began stumbling over his remarks. "The more we dive into this topic of critical minerals, the more I'm convinced that Superman isn't the only one who can be brought to . . . its knees by rare minerals," he said, laughing. Going off script, he told the room, "They were getting creative on this one." Everyone else laughed.

The rest of the hearing was more serious. Manchin argued that energy independence was crucial in the wake of Russia's invasion. He was skeptical that EVs were a real solution when so little of their supply chain was domestic. Then he parroted his usual talking points, saying China was responsible for 73 percent of lithium-ion battery production. For months, he'd been arguing with Schumer about consumer incentives for EVs. Manchin kept talking about hydrogen cars, even though just 3,000 were sold nationwide in an entire year, compared to over 700,000 EVs. His fixation on hydrogen likely stemmed from Toyota, which had a large manufacturing plant in West Virginia. The company had bet big on hydrogen and hybrids, after their successful Prius, and lagged behind on EVs.

Hasan Nazar watched the hearing live on his computer, knowing the stakes were high. The witnesses needed to convince Manchin that the EV supply chain could be onshored. As the director of federal policy for Tesla, it was Hasan's job to secure incentives that encouraged more people to choose clean cars.

Two decades earlier, on September 10, 2001, Hasan had arrived in the United States to start college. The next day, people who looked like him were suddenly hated. After this experience, he wanted to work in civil rights, trying to protect Muslim Americans, but energy policy kept calling him. After graduation, he ended up in DC working at the League of Conservation Voters, then he joined Obama's Department of Energy, before landing at a company called SolarCity. Tesla acquired it six months later, and Hasan started running their federal policy. It was a visionary company at the time. Tesla had already launched the Model S, its first mass-market EV, and would soon release the Model 3, a car affordable for everyday Americans.

But headwinds were growing at Tesla. Hasan spent his early years there trying to fix a problem from the Obama administration. As part of the auto bailout Brian Deese negotiated, each automaker agreed to electrify in the long term. In exchange, the recovery bill included a $7,500 tax credit for Americans to buy EVs. Legacy car companies didn't want any manufacturer to pull too far ahead, so they insisted the incentive only apply to the first 200,000 cars sold.

The trouble was, both Tesla and GM blew through their caps by 2018. Now the policy punished them for acting fast, as their competitors continued receiving subsidies. Hasan had come close to fixing the problem with a bill in 2019, but President Trump blocked it, believing it would benefit Americans who voted against him.

Just after the November 2020 election, Hasan could see that if Democrats swept the Georgia runoffs, a trifecta would create an opportunity to improve EV policy. But he knew that Tesla alone was unlikely to overcome opposition from fossil fuel companies and other automakers. They needed a coalition.

That winter, he helped launch the Zero Emission Transportation Association, or ZETA. To join, companies had to commit to working toward 100 percent EV sales by 2030. The founding group included electric utilities and charging companies. But as the Biden administration took office, ZETA grew, picking up mining companies that extracted minerals for EVs.

Now more than a year into Biden's presidency, Manchin still wasn't budging on supporting electric cars. This hearing was their opportunity to convince him that batteries *could* be built and recycled in the US if Congress set the right incentives. Hasan was hopeful that Manchin might finally change his mind. He'd helped ensure that a ZETA member was a witness that day: JB Staubel, from Redwood Materials. He was Tesla's former chief technology officer and now ran a battery recycling company in Nevada.

If Manchin wouldn't support a consumer incentive, Hasan thought he might support EV production farther upstream, with a manufacturing tax incentive. Manchin already had a bill for manufacturing in former coal communities, but that was for building new facilities. Hasan knew they also needed an incentive for each battery produced.

There was precedent for the idea. In the House reconciliation bill, solar advocates had secured a tax credit that gave companies a bonus for each solar component they manufactured. If they could get the same deal for batteries, each EV made in America would cost thousands less. But Manchin was still concerned that minerals were being extracted in the US, only to be processed in China, and then shipped back. This drove Hasan nuts. He knew batteries could be made in America.

Hasan brought the manufacturing idea to several Senate staffers, including Adrian. They agreed it would be a great substitute if Manchin killed consumer support for EVs. Modeling suggested it could reduce the cost of each car by $3,500. Plus, building more clean cars would create jobs across the country. Legislators always cared about jobs in their district, but manufacturing jobs held a unique allure. A person could build a career in these factories and raise a family on a good salary. If a climate law could build these jobs in Republican districts, it would be much harder to repeal.

But getting Schumer's team on board was the easy part. Hasan knew it would be much more challenging to convince Manchin. Perhaps expanding the credit to apply to critical mineral extraction farther upstream might help. And so might finding companies willing to operate in West Virginia. At the beginning of 2022, an electric school bus manufacturer announced plans to open a facility there. Form Energy, a long-duration storage company founded by a Tesla alum, also said it would consider West Virginia for its facility, given the right federal incentives. If Manchin wanted to deliver for his state, all he had to do was pass the climate bill with manufacturing incentives for batteries and critical minerals. As the hearing wrapped up, Hasan hoped that was what the senator had concluded.

With energy prices still high, negotiations over the reconciliation package started moving. By May, Adrian was meeting with Manchin's team once or twice a week. He aimed to keep the pressure low, build trust, and address their concerns. "Whatever they threw at us, we would take it seriously," Adrian told me later. "We would work through it, in excruciating detail, and just wear them out. We just kept churning through everything." There were still areas where the senators' teams couldn't agree. But with each passing week, fewer items went unresolved.

Still, Schumer and Manchin were far apart on top-line spending. Manchin wanted to cut the price tag. With inflation spiking, he regretted his vote for the American Rescue Plan even more. Injecting all that money into the economy had caused it to burn hot, leading prices to rise. Manchin's pet peeve became paying people to "stay home." When he talked to small businesses in West Virginia, they told him they couldn't find employees. He thought the federal government was to blame, with all the COVID handouts. People needed to get back to work.

While the House reconciliation bill that passed in November included $555 billion in climate investments, Manchin wanted to spend less—between $300 and $350 billion. Adrian needed to figure out how to cut spending without compromising pollution reductions too drastically. He began bringing Tim to meetings with Manchin's team, using the ambition tracker to figure out what to leave on the cutting-room floor. When the Manchin team started asking Tim how eliminating a given program would affect the emissions goal, that's when he knew they'd started seeing the world like he did.

Without the CEPP, Adrian knew the tax credits to build wind and solar projects were the most important policies in the package. First passed under Republican presidents, the production and investment tax credits drove massive growth in wind and solar for decades. The trouble was, they had to be extended every few years, creating chaos in the sector and driving up costs. Senator Ron Wyden, who chaired the Finance Committee, wanted a long-term extension. His team came up with a brilliant idea: they should continue until power sector emissions fell to 25 percent of 2022 levels. That way they could last well over a decade.

Wyden also wanted to move away from listing specific technologies and toward a technology-neutral approach. Anything that could generate power with zero emissions or store that power could qualify. This would have the benefit of covering technologies that weren't invented yet, while removing problematic ones like biomass. The trouble was, the House Ways and Means Committee didn't like this approach. After significant negotiations, Wyden got them to agree to a five-year transition period for the technology-neutral tax credits. But when the Joint Committee on Taxation estimated how much these tax credits would cost, they found that a longer transition increased spending. Adrian could see that if Schumer reneged on Wyden's deal with the House and went with a three-year transition, it would save $70 billion. The House would just have to accept it.

The other thorny issue was who could claim these credits. In the past, they were deducted from a company's federal tax liability. But as Brian had explained to Manchin, if the company didn't owe federal taxes, they'd have to partner with a bank, which would take nearly a third of the credit's value executing the transaction. This system also didn't allow nonprofits and other tax-exempt entities—like rural electric cooperatives, municipal utilities, and schools—to use the credits at all. There were two proposed solutions. The first was called "direct pay," where the government paid the credit value in cash if no taxes were owed. The second idea was "transferability"—allowing those without tax liability to transfer credits to others who owed the government taxes. The trouble was, as Brian had learned in that car ride with Manchin, the senator was skeptical of both ideas.

And then there were the problematic policies Manchin wanted to add to the bill. President Biden had kept his campaign promise to stop new leasing on federal lands for the first six months. Then, a Trump-appointed judge overturned Biden's executive order. Still, in the first 19 months, Biden offered fewer public lands for fossil fuels than any president since World War II. Although many environmental groups had forgotten, Manchin had not. And he was pissed off. His team sent over proposed text to Schumer, requiring that before any public lands could be leased for clean energy, others had to be offered for fossil fuels. When Adrian read this proposal, his heart sank. He told Gerry that if this was in the final package, it might be a deal-breaker for environmental groups. This could tank the bill.

29

DECLARATION

On Monday, June 6, President Biden declared he was using the Defense Production Act to build more solar panels. Since it would take years to ramp up domestic manufacturing, he also used his emergency powers to put a two-year pause on new tariffs for imported solar. The press release called it a "bridge" to the "American-made clean energy future." Sonia had written those words. This was her secret plan. She'd solved the solar crisis.

It had not been easy to figure out. Department of Energy staffers had done heroic work over the past two months. To get a handle on how much solar was stockpiled, they'd combed through import records and matched it against deployment data. Then they'd gathered the latest manufacturing data to understand the fraction of domestic needs US suppliers could cover. With this information, Sonia made the case inside the White House that they needed to act with urgency.

The first step was showing the Biden administration was serious about investing in solar manufacturing. Although the president had just used the Defense Production Act to deal with a baby formula shortage, there was widespread resistance within the White House to invoking it for clean energy. Sonia made the case that pausing tariffs without a plan for domestic manufacturing would anger organized labor, a key constituency. Then she argued that they shouldn't just use this crisis to build the domestic solar supply chain. They also needed to ramp up heat pumps, insulation, transformers, and clean fuels. She convinced her colleagues over several weeks. When the president invoked the Defense Production Act, all these technologies were included. Sonia knew this would help build momentum for the reconciliation bill.

Since it would take years for the solar supply chain to be brought back to America, they also authorized imports to continue for 24 months without tariffs, using the narrow emergency powers under Section 318 of the Tariff Act of 1930. This would give the Commerce Department time to conclude its investigation. In the interim, imported solar would keep flowing without accruing fees. It wasn't the broad climate emergency activists were calling on the president to invoke. But it was a solution to a real crisis. Sonia had managed to thread a complicated needle.

When I heard that the president was invoking the Defense Production Act for heat pumps, like we had been calling for since Russia invaded Ukraine, I was thrilled. By then, I was more than two years into federal activism and had yet to see a win. Knowing that something big had happened that wouldn't have occurred without our organizing was a feeling I hadn't experienced before. It felt powerful. When I reached my colleagues at Rewiring America, we celebrated.

And then, a week later, a friend at the Department of Energy called. He'd been trying to implement the president's declaration and provide funding for domestic heat pump manufacturers. "There's no money," he said. He'd asked around, and it turned out that there wasn't actually funding for any of the technologies the president had directed be manufactured in America. We had assumed some existing funds could be repurposed, but there was nothing. We'd still have to get the money from Congress. It was a hollow victory.

On top of that, life at home started to come apart at the seams. All my friends with young kids felt the same way. During the first three years of the pandemic, more than one in ten care providers vanished. In Santa Barbara, it was nearly impossible to find childcare.

After a months-long search, we hired a nanny to take care of our babies during the day. She'd been with us for a few months when she arrived one morning in June, explaining that her daughter had just come down with COVID. They'd spent time together over the weekend, so she took a test, and it was positive. We were grateful to have caught it early, hopefully before exposing the babies. I told her I would pay her sick days for as long as she needed. We thought she might be gone for ten days, but it turned into weeks. I pieced together backup childcare one day at a time, when I could even find someone.

The prospect of the twins getting COVID was always looming. We worried because their prematurity meant their lungs had not fully developed in utero. To protect them, we avoided taking them inside buildings as much as possible. That spring, when masking requirements were lifted, my kids were too young to mask themselves. The federal government had still not approved vaccines for kids under five, despite positive trial results being available for months. It felt like we were running out of time to get them vaccinated before they were exposed. Finally, on June 20, the FDA approved the vaccines. The first day the babies were eligible, my husband drove 90 minutes to get them their shots. They just needed one booster a month later. It was a huge relief.

Balancing work and two infants was brutal. My bed became an office. Most days, I woke up early because someone from the East Coast was calling or texting. I would open my laptop and just start working in my pajamas. Throughout the day, I got up to feed the kids, put them down for a nap, or read them a story. After they went to bed around seven, I would work until my head hit the pillow. And then I did it all again the next day. Days would pass when I forgot to eat, get dressed, or shower. At the time I wrote in my journal: "I have to remember to leave the bedroom—maybe even the building—on occasion!"

Pregnancy, childbirth, motherhood, and activism: all of it had started to catch up with me. My hands broke out in a rash. Then I started to have chest spasms, making it difficult to breathe. My physical therapist said my diaphragm was stuck because I spent too much time hunched over a laptop. When I came down with my third cold in three months, I realized that the costs of neglecting my physical recovery were piling up.

Instead of taking a break, I leaned further into work. While we waited to see if the climate bill would ever come up for a vote in the Senate, I turned my attention to California policy. I couldn't handle the idea that I had worked so hard with nothing to show for it. Taking on another project—while nuts in hindsight—somehow felt like the right thing to do. It was an emotional insurance policy.

At the time, Governor Newsom and the California legislature were putting together a climate package, with $50 billion in spending. Working with Rewiring America, we met Jonathan Klein. When his daughter was in seventh grade in 2019, she asked him to chaperone a youth climate strike. At the time Jonathan was a teacher, but the experience radicalized him. He decided to quit his job to start UndauntedK12, a nonprofit focused on getting fossil fuels out of schools. We began working together to try to secure state funding for school electrification. While there was already money for electric school buses, there was almost nothing to electrify school buildings, where children spend the majority of their waking hours. It was a challenging campaign, but eventually we secured $235 million for the effort, supporting schools across California to install heat pumps. I clung to that victory like a talisman.

Ricketts and I were also busy developing a plan B, if the federal bill fell apart. Although other groups were calling for a climate emergency, we were skeptical that this would give the president substantial new powers. Instead, we focused on using the Clean Air Act to set tighter regulations on pollution from power plants. We started working on another report that made the case for more ambitious regulations. The challenge was a pending case before the Supreme Court would likely limit the federal government's authority. Not knowing how the court would rule made it difficult to understand what President Biden could actually do. The longer we looked at the potential for executive action, the more paltry it seemed compared to a climate law.

For two months, Sonia and her family lived in a converted garage outside her in-laws' house in New Mexico, as she continued working long hours for the White House. But in June, they finally found a rental in the foothills of Taos, a remote town in the high desert. Their new home was at the end of a long driveway lined with rusted-out cars. Flickers and magpies flew back and forth over the road. Acequias ran across the land, forming small streams and mud puddles. On weekends, she took her toddler to splash in the water. They would occasionally find beaver lodges and hear coyotes call. Her family was happy. And Sonia was still able to work for the White House. From her office window, she could see the Taos Pueblo's bison herd.

But she was beginning to doubt that a climate law would ever pass. One night on the phone, early that summer, Sonia told me, "It's not really a negotiation. It's just a slow and painful dissection of our dreams."

30
GETTING TO YES

By mid-June, Adrian was more worried than usual. He was constantly calculating how many working days were left until Friday, August 5, when the Senate would head home for recess. Once Schumer, Manchin, Sinema, and the 47 other Democratic senators landed a deal, that was just the beginning. Committees would need to finalize the text, stitch the bill together, send it to the parliamentarian, and put it through a Byrd bath to remove any impermissible sections. Then there was the Congressional Budget Office, which needed to score every provision in the bill—a process that could take weeks for a single item. Oh, and the lead CBO energy analyst had just emailed Adrian that she'd be on vacation for the first two weeks in July. Even if they got all of this done, they would need to get the bill to the floor and make their way through the perilous vote-a-rama, which could stretch for days.

Looking at all the steps, Adrian knew they needed a deal with Manchin by early July to have a chance of passing the bill. After August, the midterm elections would consume members' time and they would never get it done. There were also healthcare policies that were about to lapse. At a certain point, Schumer might just have to pass a bill that extended those programs, jettisoning anything Manchin wouldn't agree to. To say Adrian was stressed was an understatement.

At home, things were no easier. Two kids were way more intense than one. After his wife's maternity leave ended in May, Adrian was supposed to take over looking after their baby, but there was no way he could take his paternity leave now. They'd just have to put him in daycare. The work was undermining his ability to be present with his kids or be a good partner. But he couldn't stop now.

At least they were making progress on the bill. On Friday, June 24, Manchin's team sent over an offer for $350 billion in climate investments. While he remained noncommittal on consumer rebates for EVs, he had come around on including batteries and critical minerals in the manufacturing tax credits. That alone would make EVs cheaper. But direct pay for clean energy projects continued to be a sticking point. One afternoon, Manchin was in Schumer's office for yet another meeting when the issue was mentioned. "I can't do direct pay," Manchin said. "We can't just be giving people money. I can only do tax credits." Adrian, in desperation, spoke up, saying, "What about rural electric co-ops, senator? They don't pay taxes. Since investor-owned utilities can take the credits, we could put these rural utilities at a competitive disadvantage." Manchin considered the argument. It was a fair point. "Okay," he said, "Well, maybe we can do it for the co-ops." But Adrian wasn't satisfied with that narrow victory, and piped up again. "What about other nonprofits, Senator?" Manchin stared at him again and finally replied, "Well, I'll think about it." Adrian locked eyes with one of Manchin's staffers. He knew they'd been working overtime to salvage direct pay. This was their green light to finalize a solution.

Manchin was also obsessed with permitting reform. He was frustrated that activists were delaying the Mountain Valley Pipeline, which would transport gas through his state. He began pressing Schumer on policies that made it easier to permit energy projects—clean and dirty alike. The trouble was, the parliamentarian wouldn't allow this kind of thing in a reconciliation bill. It would need to pass separately, through regular order, with Republican support. Senator Carper and Senator Schatz had begun work on this permitting reform package with Manchin's team. But Gerry and Adrian didn't know how they could assure Manchin that this side deal would eventually pass. He might just have to settle for a verbal agreement from Schumer that he would try to make it happen.

In late June, Adrian called on a Sunday afternoon, and we ended up talking for over an hour. "We've reached a point where I'm more worried about progressive votes than Manchin," Adrian said. The climate provisions were still being negotiated, but he could see the end in sight. "My big concern going forward is we are going to be dealing with a paradox,"

Adrian continued. "This will be the biggest climate bill in American history, it will be transformational, but it will also be measured against the House bill. And there will be a lot of disappointment because Build Back Better would have done even more." Six months earlier, the Schumer team estimated that the House package would deliver a 45 percent reduction in carbon pollution by 2030. But, Adrian thought the latest Senate version would probably land closer to 40 percent.

What I really wanted to know was if policies I'd worked on were still alive. In quick succession I asked about the electrification grants, state challenge grants, green bank, environmental justice investments, tax credits for heat pumps, and funding to help close coal plants. I was shocked to hear they were all still in there.

Adrian also brought me up to speed on the positive new additions, including the clean manufacturing tax credits. He didn't get into specifics but said these would be ten-year incentives. In the long run, we both knew this would change the politics. If there were jobs in Republican districts, it would be much harder for the entire party to oppose a clean energy bill down the road. "That will pay dividends," he said. I told him he should also try to get funding for heat pump manufacturing. Apart from creating more jobs, it might also help win over progressives.

Of course, it wasn't all good news. If my pet project was funding for heat pumps, Adrian's was making electric vehicles cheaper—and that was not going well. He had managed to secure $7.5 billion for charging stations through the bipartisan infrastructure law. And now Manchin was open to EV manufacturing incentives. But Adrian was struggling to find agreement on consumer rebates. "We will get something," he told me. "We'll take whatever comes from the ashes of Manchin's hatred of electric vehicles. This one hurts me personally."

Other policies were hanging on by a thread. Manchin had blocked our effort to give Americans 30 percent off a heat pump, instead capping it at $2,000 a year. Plus, he had jacked up the gas furnace incentive from $150 to $600. Senator Sanders and advocates were still pushing to make the solar tax credit more accessible to low-income Americans, but that was unlikely to happen. "It's in very bad shape," Adrian told me. "We are trying our best to salvage parts of it." He also confirmed that the $30 billion for the Civilian Climate Corps, which the Sunrise Movement had championed, was gone. Clearly, Manchin did not want to give victories to the kids protesting him on the street.

At times, it felt personal, like I was cursed. "Any time something becomes a prominent progressive campaign, he notices and he kills it," Adrian told me. "He has killed everything that you have made successful. It's like he's targeting you specifically, Leah." Of course, Manchin likely had no clue who I was. It was about what made him tick—and what ticked him off. We climate activists often believed that media attention and protests helped our cause. Our method of making things popular would work a lot better if the Democrats had a few more seats in the Senate. Ricketts and I had learned that we had to keep quiet about parts of the bill if we wanted to ensure they weren't cut. There were policies we would whisper about on phone calls, as if Manchin might somehow overhear us.

Plus, Manchin wasn't just subtracting programs—he was also adding new funding to support fossil fuels. Anyone reading the news knew that. I asked Adrian to tell me about the garbage that Manchin had shoved into the package, but he wouldn't say a word on the subject. I worked backward from what Manchin had mentioned publicly and floated educated guesses. "What about the Mountain Valley Pipeline?" I asked. "I can't for the life of me figure out how that would fit into the reconciliation rules." Adrian provided no clues. "Okay," I continued, "then what about subsidies for carbon capture and sequestration—is Manchin pushing to expand those?" Again, Adrian gave me nothing. "Anything shitty is measurable, and anything good is measurable," was all he said.

Which brought us to the key problem: how would we get this over the finish line? "You're the first person I'm calling," he said. "No one has this level of detail yet. And there's a reason for that: you know how to talk to reporters." Once a deal became public, we both worried that the media would focus on the pounds of flesh Manchin had extracted, not all the climate investments left intact. If there were enough problematic provisions, the environmental movement might turn against the bill, which could prove deadly.

Together we tried to figure out how to explain a long and complex bill that would change the world for the better—and include some bad concessions. We needed the narrative to move away from top-line numbers and tax credits toward specific stories about helping everyday Americans. We could talk about the money to buy solar panels and heat pumps. Or the funding to help struggling farmers profit from climate action. We could talk about the environmental justice investments, like billions to clean up pollution from ports.

But until the bill was on the president's desk, Manchin was still the most important audience. Stories had to hit on things he cared about, like kickstarting manufacturing in former coal communities.

Before we hung up, Adrian told me he felt hopeful. They were on track to get the bill text nailed down by the second week in July. That would leave enough time for votes before recess. Adrian felt confident they would make the deadline. "Manchin is just in a different mindset now than he was in the fall," he said. "We are past the point where he could blow things up."

31

THINGS FALL APART

After my call with Adrian, Ricketts and I got to work. One reason the Waxman-Markey bill failed was that the environmental community didn't unify behind it. Instead, the movement splintered, and over a decade passed before we got a second chance. If another climate deal fell through, we would be choosing to remain locked into fossil fuels. I wasn't willing to make that choice. Some groups were likely to oppose this bill no matter what was in it—that was how they were wired. We just needed enough groups to get behind it.

My next call was to Varshini Prakash at Sunrise. When I told her I thought a Senate bill was coming with most of the climate investments intact, she was thrilled. "Are you serious, it's going to be that good? Is it really going to happen?" she said. "I think I am going to cry. The kids are so sad." After the hunger strike, her organization had fallen into internal conflict and struggled to find its focus. The young activists had stopped believing a climate bill would ever happen. She said she was ready to rally her team, even if their signature program, the Civilian Climate Corps, was likely gone. I admired this about Varshini: she was both pragmatic and radical at the same time. The Sunrise leaders had built an organization for this moment.

But climate policy veterans were skeptical a deal would be well received. When I reached Bill McKibben, he worried that concessions to Manchin could split the movement. The most problematic additions to the bill were likely permits for the Mountain Valley Pipeline, changes to offshore drilling rules, and approval of the Willow drilling project in Alaska. No one on the outside could figure out how these compromises could make it through the reconciliation process. But we also knew Manchin wouldn't agree to a deal without some fossil fuel sweeteners.

Others simply refused to believe a deal would ever come together. During a coalition call, people started to ask who had taken the red pill or the blue pill, referencing *The Matrix*. Some activists and lobbyists had chosen to accept the painful truth that this bill would never get done, while others were still naively hoping it would happen. It was the veterans who'd lived through the Waxman-Markey days who thought we would fail, especially with just 50 votes in the Senate. They'd been burned too many times. But for me hope was still there, springing eternal.

Even though I had no details on what Manchin had added to the bill, based on Adrian's advice, I told activists that we needed to be prepared to support it, even if it was imperfect. But days later, when a journalist asked me for a comment, I fumbled. He was looking for a clean quote saying the good in the bill would outweigh the bad, and I was mealymouthed. I hadn't seen the deal with my own eyes or any modeling of its impacts. I found myself afraid of sticking my neck out and being attacked by other activists. If this was a dress rehearsal, I'd failed to perform. Once the deal was public, there'd be many more calls from journalists. I needed to get it right next time.

That June was the hottest in world history. On the last day of the month, the Supreme Court ruled that the federal government could still regulate greenhouse gases from power plants. But it limited the authority to within the fence line of the plants, disallowing regulations requiring utilities to reduce overall emissions. It was not as bad as many had feared. But despair still traveled far and wide. That same day, the Biden administration also delayed issuing a five-year plan that outlined where fossil fuel leases would be available on public lands. The delay angered Manchin.

And then it was July. My daughters' first birthday was coming up, and still no bill had passed. Late on Sunday, July 10, Senator Schumer announced he had COVID. The timing was terrible. The Senate was scheduled to return from a two-week break the next day, for its final session before the August recess. Schumer's team said he would spend the week quarantined in Brooklyn and work remotely. This meant he couldn't negotiate with Manchin in person.

The bad week only got worse. Early on Wednesday morning, a report estimated inflation had jumped 9.1 percent—the largest increase in over 40 years. By the time I woke up on the West Coast, Manchin had already issued a press release. He did say he was worried about high energy costs, suggesting an openness to finalizing the package. But later that day, he told journalists he wasn't sure he could support any new spending, given inflation. "We know what we can pass is basically the drug pricing," he said. "Is there any more we can do? I don't know." Then he questioned the budget reconciliation strategy, asking whether a bill should even be advanced on a party-line vote.

An ominous sense of déjà vu came over me. This felt like December, when everything went sideways and Manchin walked away. For the first time in months, I began to doubt we would get anything. Maybe the game was over, and we just hadn't noticed we'd lost.

To distract myself from the dark, swirling thoughts, I turned to long-neglected tasks: running errands, cooking dinners, exercising. When I talked with Sonia on the phone, she said she'd started stress baking.

That Wednesday afternoon, I biked to my community garden plot to try to calm my nervous system. Standing alone watering my plants, my mind drifting to the world's problems, a hummingbird approached. It flew within inches of my hand and delighted when I sprayed its iridescent feathers. The West was experiencing its worst drought in a thousand years, and the bird wanted water. In its own way, this tiny creature was telling me that our hotter planet was less safe than it used to be. In its own way, this hummingbird was telling me not to give up.

As I was biking home, my phone rang. It was Lisa Friedman, *The New York Times* climate reporter. I pulled over and picked up. She said she was writing yet another story about the negotiations and wanted to know where my head was at. Normally, I told journalists the package had momentum. But I couldn't maintain an optimistic front, and despair spilled out. I told Lisa I felt hopeless, like we'd been on the Manchin roller coaster for 18 months, going around endlessly, never getting off the ride. I hung up and biked the last mile home, my mind circling over the same thought: maybe Manchin was never serious.

The next day, everyone was miserable. Activists and policymakers sniped at one another over the smallest things. The ominous feeling I had was spreading. At 5:00 p.m., I wrapped up work, put the kids in the stroller, and headed out, thinking a walk might help. I wound through the neighborhood, taking in the parched landscape. The grass, long dead, was golden. The creek that ran along the path had evaporated. The climate crisis was visible everywhere I looked.

It was just after 6:00 p.m. when I pushed the stroller through the front door and Sonia texted. She sent an article that Jeff Stein and Tony Romm at *The Washington Post* had published minutes earlier. The headline read "Manchin Says He Won't Support New Climate Spending or Tax Hikes on Wealthy." I skimmed the story. It claimed negotiations between Schumer and Manchin had broken down for good, with Manchin refusing to pass any climate provisions until new inflation numbers were published in August. By then, Congress wouldn't even be in session. The journalists called it a "stunning setback."

"Well, that's a gut punch," I texted Sonia back. Within minutes, Adrian confirmed the story, saying he'd broken down. I felt the floor give way beneath me. Adrian had been in the room, negotiating with Manchin. If he was crying, it was really over. I held my kids and sobbed.

A few minutes later, my phone rang. It was Lisa Friedman again. I knew she was on deadline, chasing the story, and looking for a quick quote. I told her about all the people who had put off their parental leaves to help get this bill over the finish line. We had done it because we knew it was the most important thing we could do to protect our children. We knew the opportunity to pass a federal climate law might not come again for another decade. By that time, our kids would be teenagers, and the crisis would only have worsened. "The stakes are so high," I told her. "It's just infuriating that he is condemning our own children." The line went silent, as Lisa took down notes. "I'm so sorry, Leah," she said, with sadness, then hung up.

Her article ran within the hour. In it, Manchin's staff claimed he had not walked away. But I'd finally stopped believing him.

After the shock and sadness came anger—a feeling shared by many people across the country. "Senator Manchin has condemned his own grandchildren to a broken planet," I posted that night. "His actions will be recorded in the fossil record for centuries to come. What sounds like hyperbole is sadly a fact. After stringing along his Democratic colleagues for 18 months, and delaying the Biden administration's executive actions, Manchin has torched any climate deal. He does not care about the American people or the planet his grandchildren will grow up on."

Thousands shared similar messages, denouncing Manchin's inhumanity. John Podesta even joined in. An alumnus of both the Clinton and Obama administrations, he was not the kind of person to criticize a sitting Democratic senator. But he cared deeply about the climate crisis and had vocally supported the bill. "It seems odd that Senator Manchin would choose as his legacy to be the one man who single-handedly doomed humanity," he wrote that night. Although I didn't know it at the time, I would later learn that many contacted Manchin directly to express their outrage.

Lori Lodes, who ran the organization Climate Power and worked with John and Christy, pointed out Manchin's relationship with big polluters. In the past year, he'd received more campaign contributions from the fossil fuel industry than any other congressperson—nearly twice as much as any Republican. A recent investigation from journalists Christopher Flavelle and Julie Tate had found that Manchin had earned over half a million dollars each year from his coal company for the past decade. It was almost triple his Senate salary. In that moment, this was the only explanation that felt true: Manchin was bought and paid for by the fossil fuel industry.

When I finally reached Adrian, he told me what had happened. "I am a wreck," he said. "We threw everything at him in the last 24 hours." The Schumer team had essentially clicked "accept all suggested changes" in the draft bill with Manchin. Yet he still couldn't get to yes. That was why Schumer had told the world it was over.

Later that night, Sonia called. For the past 18 months, we'd been like sisters marooned on a boat, clinging to our young children, frantically paddling toward shore. We kept believing we would eventually make it, only to find that the ocean never ended. "This is just a bump in the road. We are going to win," she said, trying to convince herself as much as me. "But, you know . . . it's a really big bump." I could hear her son in the background. He was mimicking her, as toddlers do, yelling, "It's a really big bump," over and over again. We laughed between our tears.

The evening dragged on, filled with texts and calls from people I'd campaigned alongside. We were grieving together collectively, in time zones spanning the country. Suddenly, all the small differences in strategy seemed unimportant. These people were my sisters and brothers, my cousins. They understood the stakes and were devoting their lives to ensuring we had a livable planet. In that moment, I loved them all.

People sent notes, just to say thank you for the effort. And I found myself doing the same thing. I guess that's what you do after a long fight, whether you win or lose: you say thank you.

An hour and a half after the news broke, Tim Ryder from Schumer's office texted, near midnight his time. "Today is a sad day, my friend, but I want you to know how much I dearly appreciate your incredible work that got us this far. I won't call, it is late, but I have learned so much from you in this process and we clearly have a lot more work to do. Thank you for everything. Amidst the anger and sadness I am grateful for you and your insight, Leah."

"That means so much," I replied. "So sorry for how hard you worked and how painful this is."

"The work was worth it," Tim said. "We are slowly convincing leaders and I am not ready to stop fighting for that. I hope whatever you do tonight brings you some solace."

I told him I was going to cry and write.

Despite the grief, I stayed up late, well past midnight. I wrote that I'd spent the past year dissecting Manchin's every word on climate change. I wrote about the young people devastated by this news. And I wrote about that tiny hummingbird in the garden who beckoned me to give it water, the bird who asked me not to give up.

Before closing my computer, I sent the draft to *The New York Times*. With the most readers in the country, I hoped they would take the article. If Manchin was going to torch the planet, at least everyone would know who had lit the match.

32

THE LONGEST TWO WEEKS

Day 1, After Death—Friday, July 15—I slept fitfully, waking in the middle of the night with a surreal feeling so familiar to the grieving. Was this really happening? By 6:00 a.m., the news cycle was in full swing, so I got up and kept working. *The New York Times* editor had written back and said she would take the piece. When that happens, it's a manic process, with turnarounds on edits needed within the hour. And since no one knew Manchin would blow up the negotiations on a Thursday in mid-July, I had a full day of meetings already scheduled. That was probably for the best: maybe I could outrun heartbreak.

That morning, Manchin went on a West Virginia radio show to defend himself. He claimed he hadn't walked away. The editor began pressing me on whether negotiations were truly over.

I tried to reach Adrian, who was dealing with his own mixture of overwhelming work and grief. When I finally got him on the phone, he walked me through what had happened. After weeks of endless back and forth, Manchin had told Schumer: "I'm no longer on your timeline." He said the bill didn't need to pass before the August recess. Schumer knew that wasn't true—they had to act now. When he offered to make all of Manchin's desired changes and he still wouldn't move forward, Schumer decided it was over. "I desperately want to believe that there is still an opportunity on climate," Adrian told me. "I want this to happen with every fiber of my being."

Day 2, A.D.—That Saturday, my article ran on the front page of *The New York Times* website, under the headline "What Joe Manchin Cost Us." It stayed there for two days straight because people kept reading it. I spent my weekend fielding right-wing attacks at a level I'd never experienced before. Thousands of comments calling me an idiot and insisting climate change wasn't real. Many were exactly the same and clearly posted by bots. The worst ones said my children should be seized by child protective services. It was stressful, but it was also a sign: the article was effective.

Day 3, A.D.—Sonia spent her weekend working with her White House colleagues to use the full force of President Biden's executive authority. They had been holding back for 18 months waiting on Manchin. Now they could act. An event was organized for the president that Wednesday in Massachusetts. He would stand next to a retired coal plant that was now manufacturing cables for offshore wind, and declare a climate emergency.

It was Sonia's job to help prepare the declaration. When she read the first draft from the White House lawyers, it sounded like legalese that failed to meet the moment. Over the next four hours, Sonia and her colleagues transformed it into a rousing call to action. They made lists of the tools they could use to cut pollution and piss off Manchin at the same time. Biden was ready to be bold. They had everything they needed. Everything except legislation. "Paradoxically, we felt like we were the most powerful and the least powerful people in the world at the same time," Sonia told me later.

Late that night, Ricketts was getting ice cream with his family on the Jersey Shore when he got a call from a Connecticut area code. He didn't recognize the number but picked it up anyway. It was Luke Bassett, his old friend who worked for Manchin. Luke said they needed to talk. Ricketts, surprised, started to reassure him, saying, "Hey, man, I appreciate all that you tried to do here." But Luke cut him off: "I just want you to know, we are still trying." He wanted Ricketts to moderate his tone and stop attacking his boss. "How could you possibly ask me to do that?" Ricketts replied. "I get what you're saying, but what assurance can you give me that your boss is ever going to vote for anything?" Luke told him, "I can't say more. Just know we are trying." After Ricketts hung up, he concluded Luke was simply looking to make himself feel better. But he took the advice, softening his tone the rest of the week. In the back of his mind, he started to wonder if maybe Manchin would come around.

Day 4, A.D.—By Monday, Europe was in the grip of a brutal heat wave. People were dying in Spain and Portugal, temperature records were being smashed in the UK, and Manchin was the villain to blame—whether or not that was fair. I did TV segments and radio appearances, taking endless calls from journalists. After talking through what Luke had shared, Ricketts and I started ginning up media stories about how Manchin's decision to walk away had screwed over West Virginia. The bill included funding for manufacturing in coal communities and support for miners with black lung disease. If it never passed, Manchin would fail to deliver these dollars to his community.

Similarly, Hasan Nazar was told to put mining companies in front of Manchin, and he got to work organizing calls to the senator. Bill Gates also called regularly, trying to get Manchin back to the negotiating table. They had met in person just ten days earlier, and the billionaire had built a good relationship with him. Gates was an empathetic listener, who felt Manchin had been put through hell over the past six months. He tried to reassure the senator that funding for clean energy wouldn't add to inflation.

Meanwhile, Senator Mitch McConnell and Senate Republicans were ready to pass the CHIPS bill, a bipartisan effort to boost domestic semiconductor manufacturing. For the past three months, they'd held it hostage in an attempt to stop reconciliation. Believing they'd succeeded, they were now eager to get the deal done.

By the evening, the White House floated the climate emergency in the press, saying the president would announce it on Wednesday in Massachusetts.

Day 5, A.D.—Manchin told reporters he didn't want the president to declare a climate emergency, saying, "Congress should act." As if he hadn't spent 18 months blocking that action.

Sonia and Jahi were busy preparing the president's remarks when they heard that someone up the chain had called off the climate emergency. They didn't know what to make of it. Sonia heard that climate hawks in Congress were furious the White House was backing down. She hoped this shift meant that negotiations were still ongoing. But when she talked to Tim in Schumer's office, he said they were not. The speech had been changed to say that Biden would give Congress two more weeks to act. But later in the day, that language got cut. These mixed signals confused Sonia. In moments like these, when they had spent an entire weekend working on something, only to see it washed away in the river of politics, Jahi would turn to Sonia and joke: "It's an honor to serve." Gallows humor kept them sane. Sonia told Jahi they needed to keep preparing either way, saying: "No one will remember this time by anything but the outcome. We have to stay focused on that."

Day 6, A.D.—President Biden stood at a power plant near the confluence of the Taunton and Lee Rivers and said the word *emergency* five times. Then the press asked a few questions and everyone went on with their lives. Biden hadn't actually done anything. It reminded me of the British political satire *The Thick of It*. Manchin had once again made the president look like a fool.

Gina was there in person. She'd seen the speech change half a dozen times over the last 48 hours. What started out as a barn burner was watered down to a meaningless lecture. She wasn't surprised. After many decades in government, Gina did not believe the bill was lost. It would have taken Manchin proclaiming "Over my dead body" on the Senate floor for Gina to believe it was really over. And when the president winked at her that day, she knew her instinct was correct. Behind the scenes, there must still be movement. It was the only thing that made sense.

That night, Sonia sat down on her son's bed to read him a story. He was almost two years old and loved their evening ritual. He chose the book *Maybe* by Kobi Yamada. Sonia opened the first page and began reading. "Do everything with love. Follow your heart and see where it leads you," she said aloud. "Maybe you will speak up for those who can't speak for themselves . . . Maybe you are here to help in ways that only you can." Sonia was crying, but pressed on. "At times it will feel really hard. And you might make a mess of things . . . You may fail. But you will also get back up, and you will rise a little stronger and a little taller. Because there really is more inside you than you know. And this world needs your gifts, your talents, your big ideas. And maybe you are just getting started." Before going to bed, she sent me photos of these pages. They made me cry too.

Day 7, A.D.—Three senators—Hickenlooper, Wyden, and Coons—banded together as the "Never Say Die" caucus. Senator John Hickenlooper from Colorado, who shared Manchin's all-of-the-above energy strategy, tried to convince him to finish the bill. He reminded Manchin he'd poured years of effort into it. And, Hickenlooper argued, it would be counter-inflationary and create jobs in West Virginia. But Manchin didn't seem to listen. Meanwhile, Senator Ron Wyden, chair of the Finance Committee, went into salvage mode, looking for any pieces that Manchin might agree to in the slimmed-down healthcare package headed to the floor before the recess.

Senator Chris Coons from Delaware felt he was making more progress. He helped organize a group of economists whose analyses showed the bill would cut the deficit. Then Coons approached Manchin directly, and said, "There are folks in our party who are saying all sorts of terrible things about you, who believe you were stringing us along for a year and that you were never going to come to a deal because of your state or because of your conflicts of interest." Then Coons made his closing argument: "I can't think of a better way for you to prove them all wrong than to sign off on a bold climate deal. Prove every critic wrong." According to reports, Manchin reflected on the comment, then responded, "It would be like hitting a homer in the bottom of the ninth, wouldn't it?"

Day 8, A.D.—A week had passed since the bill's death. I texted Adrian that morning: "I feel sad again . . . it seems even more dead today." A few hours later, he replied: "I've binge watched *Star Wars* since last week's collapse. Felt weirdly appropriate." Adrian's kids were sick, and he was exhausted.

That night, I called Ricketts. His four-year-old son picked up. I asked if he had heard his dad talk about a man named Joe Manchin, saying he might just turn out to be a bad guy after all. "You mean like Scar and Jafar?" his son asked, comparing the senator to Disney cartoon villains. "Well, not exactly the same," I told him. "But yeah. Kind of like Scar." He passed the phone to his father. Ricketts's usual manic opening line—"What you got? What you got?"—didn't come up. Nor did his periodic postmortem analysis. We'd tried every play. We'd lost. There was nothing else to say. We'd been staring into the abyss for over a week, and it wasn't getting any easier.

Day 9, A.D.—Over the weekend, I had extra time to think. I found my mind wandering to all the parts of the bill that wouldn't happen anymore. The programs for low-income families to get solar or a heat pump. The money to help retire coal plants in rural communities. The tax credits for wind and solar. Everything we'd worked for was gone.

That day, a large group of Sunrise activists showed up at Ron Klain's house. They brought little kids dressed up as firefighters with signs that said "Declare a Climate Emergency!" Ron found the whole thing counterproductive. He was already working as hard as he could, and these protestors couldn't make him work harder. His wife went outside to talk with the group and brought their handwritten notes into the house. This only frustrated Ron further. He thought the best thing to do was just ignore them, not feed the wildlife. His wife decided the best thing to do was show support. The protestors stayed all day, then slept in front of their house overnight, and his wife, who had worked on environmental issues for decades, spent much of the evening outside with them.

Day 10, A.D.—On Sunday, my mind traced the question that failure brings: should we have even tried? It felt like I'd time traveled back to October, when Manchin killed our clean electricity plan. I was embarrassed that we'd failed again.

The only bright spot was that Ady Barkan had reached out after reading my op-ed, and I met him at his home that night. Ady was suffering from ALS and had become a prominent healthcare activist. He was living a parallel life to mine in Santa Barbara. Somehow, we both worked to hold federal politicians accountable from a sunny enclave three time zones away. We'd both used the Democratic primary to push for progress. He'd managed to interview most of the candidates for president, getting many to travel to see him in Santa Barbara.

Sitting in his backyard that warm summer evening, as dusk turned to dark, we talked about our disappointment. Ady had been trying to get home healthcare aides for people like him included in the package, but Manchin had quashed that idea months ago. By then, Ady was unable to speak with his own voice, so he used a computer to share his thoughts. I have a tendency to fill awkward silences, but that doesn't work when you are waiting for someone to respond by using their eyes to spell out each word. He told me that I reminded him of how he was before he got sick—bursting with energy to change the world. In truth, he was still that person.

Day 11, A.D.—On Monday evening, I texted Adrian again. We'd been talking with Senator Hickenlooper's staff, who wanted Schumer to hold the reconciliation bill. They thought Manchin would come back to the table if he was just given enough time. Adrian wrote back: "It's funny because these Hickenlooper staff have never reached out directly and I keep hearing this from other people." Later that day, Manchin's team announced he'd come down with COVID and would be working remotely.

Day 12, A.D.—By Tuesday afternoon, a close colleague was feeling hopeful. He thought there'd been progress over the weekend. I thought he was delusional—smoking too much hope-ium. But I passed the intel along to Adrian anyway. He never responded.

Day 13, A.D.—On Wednesday morning, the Senate passed the bipartisan CHIPS Act, 64 to 33. Manchin was still quarantined, so he didn't vote for it, but would later tell the press he supported the bill.

In the past two weeks I'd aged years, gray hairs sprouting from my temples like weeds. That afternoon, just before the close of business on the East Coast, I sat down to read my children their favorite book. I now had more free afternoons with them—one bright spot in an otherwise bleak period. We'd read this particular book so many times the pages were worn, and I'd memorized all the words. "Every day, everywhere, babies are born . . . every day, everywhere, babies are fed . . . every day, everywhere, babies are loved." This wasn't a work of fiction—these were facts. Painful facts given the state of negotiations in Congress. Babies were being born every day, and it looked like we'd fail them again. We were leaving them without a safe world to grow up in.

I chanted the book's words aloud, like an incantation: "Every day, everywhere babies are loved—for trying so hard, for traveling so far, for being so wonderful just as they are." At the end of the book, the baby turns one year old, a milestone that was just days away for my own children. They could now sit up on their own and listen to the story. Before long, they'd be walking.

Just after I finished reading, my phone lit up with a message from Adrian. He sent the celebration emoji, without any other context. I was confused. Why was he telling me to celebrate?

Seconds later, Sonia sent me a link. It was a press release that had just gone live on Manchin's website, titled "Manchin Supports Inflation Reduction Act of 2022."

"What the hell is the Inflation Reduction Act?" I yelled.

I sent the link to Adrian, asking, "What does this mean?"

He replied: "Get ready to roll."

33

"THE INFLATION REDUCTION ACT"

Weeks earlier, when it was still the middle of July, Gerry found himself more worried than usual. Although he worked with Manchin's staff most days, they weren't making progress. Manchin's bottom line just kept shifting. With time running out, Gerry's anxiety grew.

His colleague Anna Taylor was even more frustrated the bill wasn't done because she was nine months pregnant. As Schumer's principal tax advisor, she played a crucial role in negotiating the package, putting in long days as her belly swelled. By then, Anna had worked for Schumer for a decade. Originally from the cotton and rice fields of eastern Arkansas, she'd gotten into politics as a teenager because of Blanche Lincoln. Watching Lincoln on TV when she first ran for Congress in 1992, Anna had fallen in love with this confident woman and decided she would work for her one day. Just six years later, Lincoln was the youngest woman ever elected to the Senate. Anna still dreamed of joining her office and decided to stay in state for college, knowing that landing a job in Arkansas politics would require her to stick to her roots.

By some miracle, Anna got a job working on tax policy with Senator Lincoln right out of law school. Even after Lincoln eventually lost reelection, they stayed close, becoming good friends. Lincoln even threw Anna a baby shower that summer, and Gerry came. They feasted on crabs outdoors on Gibson Island and toasted her future baby. Her pregnancy had been long and hard, full of fear that her daughter might not live due to a rare genetic disorder. But the tests had come back clear, and Anna was hopeful. She was excited to be a mom.

But weeks after the baby shower, the bill still wasn't done. And Anna started to worry it wouldn't be finished before she went into labor—if it ever got finished at all.

On Thursday, July 14, Gerry and Schumer's chief of staff summoned Adrian, Tim, Anna, and a handful of others to a meeting. It was already late in the day, and Adrian had driven home. As he listened in, he could see his kids playing through the glass door separating his office from the living room. Gerry was quiet. He'd asked his boss to break the news because he knew if he did, he'd start crying.

Three hours earlier, Manchin's chief sent Gerry a text: "I've got to come see you right away." Gerry hoped this meant they had a final deal so they could meet the deadline. When Manchin's chief arrived a few minutes later, Gerry led him into Schumer's office for a private conversation. The senators had already gone home for the weekend, so they were alone among the office's Roosevelt portraits. When he was in town, Schumer insisted on lighting the fireplace most days. He thought it gave the room a nice ambiance. But that afternoon, the chimney was cold.

After the door closed behind him, Manchin's chief said simply: "We can't do it. We've got to wait until probably September."

Gerry felt his blood rising. He tried to maintain calm when negotiations got tense. That was his job. But at that moment, he lost it. "What the fuck," he said. "We've got to do this before the August recess or else you know we're not going to get this done." Manchin's team had poured infinite effort into this bill, and Gerry knew they wanted to finish it. He tried to reason: "All the work that you, me, and all of our teams have done for years getting ready for this moment, if you do this, it's all going to blow away." Gerry knew he'd have to call up Schumer, and then it would be over. They'd decide to just move a smaller package. "I know for a fact, with 99.8 percent certainty, that is what's going to happen next if you don't change your mind," he said.

Manchin's chief couldn't even disagree. "I know that," he said.

"Well, why don't we try harder?" Gerry yelled. "Go fucking call Manchin. You want me to talk to Manchin? I'll call Manchin."

"No. You either wait until September or we do prescription drugs now. It's your choice," he replied.

"That isn't a choice," Gerry said. "Because September is too close to the fucking election. Or someone could die and then we would pass nothing. We have to do it now before the August recess." Even as he said these words, he knew they wouldn't change anything. The reality of the situation hit. "I do not want this," he pleaded, becoming emotional. "I've spent my career getting ready for this moment. Do not do this, man. Come on, let's figure it out."

"No, this is it," Manchin's chief said, and walked out of the office.

The rest of the day unfolded as Gerry predicted. Schumer called up Manchin, pressing him on what his chief had said. Manchin wouldn't commit to passing the energy and climate investments before the August recess. So Schumer and his chief decided they had to pass a smaller package, with just the tax and healthcare provisions, and jettison the climate funding. Time had simply run short. Just three hours later, Gerry faced the brutal task of informing the rest of the team.

"I have some bad news to share," Schumer's chief told the team. "We're not going to do the climate package." Tim, who often texted Adrian during calls, sent him a message. "Is this real? I don't understand what's happening."

Adrian felt like he'd just heard that a close family member had died. He couldn't accept it. Faced with a deep well of grief, Adrian chose to bargain. Breaking rank, he told the chief and Gerry that abandoning the climate investments was not an option. "We have to keep trying," he pleaded. "We can wait. If Manchin said he was open to waiting for the new inflation numbers in mid-August, we can just wait until then. We should make the senators stick around. We just need a little more time."

Gerry cut him off. "The decision is final," he said. "I'm upset too, but we can't do anything about it." Adrian would have to accept it. Schumer and his senior team had concluded that Manchin had been stringing them along, and it was time to call his bluff. They could not abandon the progress on prescription drugs and taxes, and they needed to extend the healthcare subsidies before they expired. They had to move on. Even though Gerry felt shattered, he didn't have time to dwell in his feelings. Fifty more things needed to happen that night: they had to call Speaker Pelosi, tell the White House, then figure out how to break the news to the world without Schumer getting the blame.

Anna, still in shock, spoke up. "We should still feel very proud of this work," she said, emotional. "In this town, nothing ever really goes away. It will get used, it's just a question of when." But no one seemed to hear what she was saying. In that moment, they didn't want a cheerleader, and Anna was too depressed to be one.

When he hung up the call, Adrian began to sob uncontrollably. Something inside him snapped. He felt like he'd let his children down in one of the worst ways imaginable. That this failure would leave them with a potentially uninhabitable planet.

He spent the next three days in a dissociated state, barely sleeping. Schumer was worried about him and called several times, telling him to take a few days off—albeit after Adrian helped Schumer debrief climate advocates, explaining what happened. Adrian's wife watched as he crumbled. She'd never seen him that upset before, not about anything. She felt terrible for him and tried to understand the depths of his grief, but they still had two little kids to take care of. For the next several days, she felt like a single mother.

Adrian saw the outpouring of anger at Manchin online, but he couldn't focus on it. He couldn't focus on anything. That night, unable to sleep, he started watching the *Star Wars* films in chronological order. Over the next three days, he got through all nine films. Something about watching Anakin Skywalker turn into Darth Vader felt appropriate.

On Sunday morning, he drove into the city for brunch with one of Manchin's staffers—his friend, ironically named Luke. Adrian brought the baby along, to lighten his wife's load at home. He found Luke Bassett in a weird headspace. "We didn't walk away from the negotiating table," his friend said. "We're still working. I've been working all weekend."

"But Schumer's team isn't working," Adrian responded. It didn't make any sense.

The next day, Adrian went back to the office. It was Monday, and the Republicans were now pushing to get the bipartisan CHIPS bill done before the August recess. He'd put a lot of time into that package and wanted it to pass—even if he was heartbroken over the climate bill. At the end of the workday, Gerry sent Adrian and Anna an email, asking, "Can you guys meet ASAP?" Less than half an hour later, they were once again on a Zoom with the chief. But this time, they had good news.

Gerry told Adrian and Anna he'd received a text from Manchin's chief of staff that morning, asking to meet as soon as possible. Gerry immediately knew what it must be about. Right around the same time, Brian Deese's phone rang. It was Senator Manchin. "You know, I never thought this deal was dead," he told Brian. "I want you to come up to the Hill this morning to meet with my staff. Can you make time for that this morning?" It's not like Brian would need to check his calendar. There wasn't anything higher priority than getting this bill done.

Cautiously optimistic, both Brian and Gerry went down to Manchin's hideaway, deep in the bowels of the Senate. Senators used these rooms while waiting on votes, and each reflected their personality: Manchin's was plastered with photographs of West Virginia and had a bar to entertain guests, something he loved to do.

When Gerry walked in that morning, an ominous feeling of déjà vu set in. He'd been there one year earlier, drinking whiskey with Manchin and Sinema as they celebrated a handshake deal on the climate package. It had turned out that framework wasn't worth the paper it was written on. But that morning, Manchin's chief of staff was not there to serve Gerry or Brian a drink. Instead, he came with three copies of a one-page document, which he passed across the long table. The chief explained that he'd sat down with his boss over the weekend to work out this proposal. Manchin didn't want to end up the villain. As he read the proposal, Gerry found his anger rising. "What the fuck," he began to yell. "This is basically where we were three days ago!" Brian, sitting in the middle of the table, cut him off. "Gerry—you're right. This is exactly what we were talking about. If they're saying this is what they can do, why are you pissed off? This is great!"

Gerry knew Brian was right—but only if they got the deal done. "I've been sticking my neck out, defending you guys, saying that you were going to fucking do something here, for a year. I'm willing to do it one more time, but it's got to be before the August recess, and this has got to be it," Gerry explained. "No more fucking around. This is the deal. We're locking arms and doing this together."

Manchin's chief looked Gerry dead in the eye and told him, "This is the plan. We're on."

Later that day, Schumer returned to DC and headed down to the basement below the Senate chamber. He was going to talk with Manchin himself to finalize the plan. When he found the room where they would be meeting, he remarked that he didn't even know this place existed. "It's your room," his staff explained—one of many spaces the majority leader controlled across the Capitol. Manchin was scheduled to arrive later, so no one would suspect he was meeting with Schumer.

"Are you still upset?" Manchin asked when he walked in.

"I'm very discouraged," Schumer replied.

"Well, you shouldn't be because something positive could be done if we all want to work rationally," Manchin continued. "This is ridiculous. We can recalibrate, see if something can be done. Can we work together and try to put together a bill?"

"As long as we finish in August," Schumer said.

The two senators left with a handshake agreement and a decision to keep these final negotiations secret. No one wanted to get people's hopes up, yet again, in case things fell apart.

Gerry told Adrian and Anna all of this, then read them the riot act. "You cannot tell a soul," he said, gravely. "No one else in the office can know outside this circle. You cannot even tell your husband or wife." After the heartbreak of the last few days, they both struggled to believe the bill was back on, let alone that it would pass.

The next day, they met with Manchin's staffers who were in on the secret. After debating the most confidential location, they settled on an Energy Committee conference room. "Let them do the talking," Gerry said to Adrian and Anna as they made their way over. After an hour of listening, Schumer's team left and walked the long hallway to the elevators in silence. Gerry pushed the button, and thankfully the next elevator arrived empty. No one said a word until the doors shut behind them. "Holy shit," Anna said. "It sounds like they're for real this time." They could all sense that the tone of the negotiations had changed dramatically. Before, Anna found herself dreading meetings with Manchin's team, never knowing what new challenge she would have to spend the next 48 hours trying to solve. Now it felt like they were all working toward the same goal.

And so began the strangest period of their working lives. Lying to everyone they knew was not easy—especially since they needed excuses for what they were working on. "There was nothing normal about the period of time that we were in a blackout," Adrian told me years later. "The degree of secrecy was beyond anything I've ever experienced and will ever experience—by far. It was very intense." Adrian's phone was constantly full of text messages from colleagues in the climate movement, who were all depressed. He felt terrible that he had to pretend he was just as sad as they were. When Anna had lunch with a good friend, she had to pose as crestfallen the entire meal. It was even harder to lie to her colleagues on the Finance Committee, who were drowning their sorrows at happy hour. At least her pregnancy gave her an excuse for not joining in. "Just think about how busy we'd all be right now if Manchin hadn't blown everything up," one close colleague told her. "Yeah," Anna agreed, like she hadn't stayed up past midnight doing their work for them.

The secrecy wasn't just emotionally challenging: it was also impractical. Usually, they relied on a network of committee staff and colleagues to hold parts of the package. Now Adrian and Anna had to do everything themselves. Digging through his files, Adrian pulled up what he thought was the latest text for the tax title. The Finance Committee had a complex numbering system to ensure they always used the newest version as the language went through endless tweaks. But Adrian could never keep the numbering straight, and he couldn't ask anyone if he had the right version. So he just sent what he had to Manchin's team, hoping for the best.

Their challenges only mounted on Wednesday night, when the Joint Committee on Taxation, which provides estimates for bill spending, sent over a new score. It was way higher than their past estimate. Something had gone truly haywire, and Anna and Adrian needed to fix it. But not before reading through the changes to the finance title that Manchin's staff had returned. They were working long hours and juggling too many documents. They were bound to make mistakes.

Adrian was thrilled when Gerry allowed him to bring Tim into the inner circle the next day. Manchin's team had been asking how much pollution the latest version would cut. They needed Tim to run the changes through the ambition tracker. When Tim found out what was happening, the past three days finally made sense. Adrian's behavior had been erratic:

he'd gone from super sad to working insane hours. Ever the optimist, the bill falling apart had shattered Tim. He was thrilled to be working on it again, and particularly happy that Manchin's team, initially skeptical of the ambition tracker, was now bought in. As they discussed the last round of cuts, Manchin's team started asking, "But what will that do to the emissions?" That's when Tim knew all his effort had paid off.

Even with the extra help, they still had one big problem: the price tag was too high. Manchin was unwilling to spend more than $370 billion and insisted on cutting the deficit. They had added a lot to the bill over the past few weeks, including money for American battery manufacturing. Manchin had also agreed on some funding to allow the president to use the Defense Production Act to manufacture heat pumps and mine critical minerals. The latest estimates were above Manchin's limit.

And there was still disagreement about funding for electric vehicles. Manchin, now bought into the idea that EVs *could* be made in America, was insisting that they *must*. Adrian developed a complex formula for the consumer credit that phased in over time based on how much of the car's components were American-made. It was clear to him that this would ultimately render the credit useless, but the independent analysts who scored the bill didn't see it that way. They guessed a lot of people would still be able to use it in the outer years. Over the next few days, Adrian put together analyses to show why the EV credits would spend less money. He explained that there were basically no cars available for purchase that met these stringent requirements. Normally, it was hard to convince these independent analysts, but this time they relented and lowered the estimate. This provided headroom to get everything else they needed into the package and still meet Manchin's demand. Like the old marketing technique, the final total was $369.75 billion—as if it was a shirt on sale, and not a massive federal bill.

A week into the secret talks, Manchin tested positive for COVID, making his final meeting with Schumer a virtual handshake. When Manchin made an offhand remark that they could call the bill the Inflation Reduction Act, Schumer pounced on the idea. "That's brilliant," he said.

On the morning of Wednesday, July 27, the press teams from both Schumer and Manchin's office were looped in. It was a busy day, with the bipartisan CHIPS Act also getting its final floor vote. Manchin wanted to post the press release on his website so he could control the messaging. They agreed to do it just before the end of the workday.

As the hours passed, Anna couldn't shake the feeling that Manchin was going to get cold feet and back out at the last minute. It was only once the press release went live at 4:55 p.m. that she felt like she could breathe again.

34

THE FIGHT OF ENACTMENT

Minutes later, Tim called Sonia. "It's back. The bill is back." he told her. Sonia was confused. "You can't mess with me on something like this, Tim. If you're messing with me, I am never going to forgive you."

"Trust me," he said. "I wouldn't do that to you. It really is back on. Go to Manchin's website, and you'll see a press release that just went live."

"Okay, so how did they mess up the bill?" Sonia demanded.

"No," Tim reassured her. "It's almost the exact same bill we were discussing before it died two weeks ago. Maybe one or two things changed. But it's essentially the same."

Sonia had trouble believing him. Why would Manchin walk away, only to return to the same deal days later? It didn't make sense. She went to the website and saw Tim was right. There was a new press release. "The Inflation Reduction Act? Manchin's Inflation Reduction Act? Are you sure, Tim? This is the same bill?"

"Yes, Sonia. It just got a new name. This is the bill. We're doing it."

Two minutes later, I was reading that same document and similarly struggling to understand what was happening. Within the hour, Schumer posted a four-page summary and that's when I knew the bill was back. It had all the things I'd worked on: state funding for building electrification, tax credits for heat pumps, grants for carbon reduction. It even included funding to manufacture heat pumps in America through the Defense Production Act. And it was full of the pieces that thousands of others, including Sonia and Adrian, had poured so much time into: tax credits for clean energy, manufacturing investments, tens of billions for environmental justice. It was like uncovering sparkling gems buried in the earth. There they were, still intact, gleaming.

Figuring out the bad provisions that Manchin had shoved into the bill was more complicated. Two emerged as the most problematic. First, Manchin had insisted on a bizarre requirement that if any public lands and waters were auctioned for clean energy, areas also had to be offered for fossil fuels. It was a bad policy that I disagreed with. But just because a parcel was auctioned did not mean anyone would bid. Over the past decade, just 12 percent of areas auctioned were developed for fossil fuel extraction. And the bill contained some good provisions on fossil fuel leasing, including an increased royalty rate and discouragement of speculators.

Second, Manchin was insisting on an additional permitting reform bill to be passed later through regular order. It would include approval of the Mountain Valley Pipeline, which ran through West Virginia and which front-line groups had been fighting for years. There was no guarantee that this second package would pass, but many activists were angry it was even on the table.

I knew the main question journalists would ask: "Does the good outweigh the bad?" They wanted to know, from a carbon pollution perspective, if this bill put us on the right path. It was abundantly clear to me that it would—the bill had at least 20 times more good investments than bad ones.

Ricketts knew we had to work quickly to shape the media narrative. That night, we made a list of 25 news outlets, and called each one to tell them this was the climate bill we'd been waiting for. The next day passed in a blur of interviews. Meanwhile, Manchin went into sales mode, even pushing the bill on Fox News. "The Inflation Reduction Act is a red, white, and blue bill. This is a great bill," he said on air. If someone had told me two weeks earlier that I'd be watching Manchin sell a climate package on TV, I'd call them delusional. This was the guy who shot the last climate bill with a gun. But there he was, sticking to his word.

Perhaps assuming anything Manchin supported must be bad, parts of the climate and environmental justice movements began turning against the package. As Adrian had predicted back in June, there was a fixation on the relatively few bad provisions Manchin had added. On Friday, two days after the bill reappeared, the journalist Mehdi Hasan interviewed me on TV. After I explained what was in the bill and how it would help bring down costs for Americans, Mehdi asked me about the bad parts. "There is also a piece of this bill that ties the leasing of public waters for offshore wind projects to the leasing of public lands for new oil and gas extraction, something obviously put in there to keep Joe Manchin on board," he said. "That provision [has been called] 'a climate suicide pact.' Do you think that compromise is worth it?" It was jarring to be asked if the bill was a suicide pact on national TV, but I stuck to my message. "I do think the compromise is worth it," I responded. "Look, it's not the bill that I would've written, it's a compromise. We have to get 50 senators to vote for this thing. And we have to be clear-eyed. We can't miss the forest for the trees here because, literally, the forest will burn down if we do." The bill wasn't perfect, but it was much better than the status quo. I'd done my best to explain the tradeoffs.

Later that evening, we were in the backyard when I noticed one of my daughters seemed tired and confused. "Something is off," I said to my husband. And as soon as those words came out, we both realized we needed to test her for COVID. It came back positive. We'd worked so hard to protect our vulnerable daughters in their first year: keeping them outside of buildings, limiting who came over, and getting them vaccinated the very first day we could. They'd had their second shots just ten days before. With their first birthday a mere 48 hours away, the timing was brutal. We canceled their party, and I kept working while taking care of two sick babies and falling ill myself. This bill was too important. I knew I would never forgive myself if I didn't do everything I could to help it pass.

And that's when I started facing attacks from a handful of activists. Some claimed I was in the basement during the secret negotiations with Manchin, green-lighting the Mountain Valley Pipeline. Another said I didn't care about people because I'd written about a hummingbird in my op-ed. One person alleged I was a "white supremacist" because I lived in Santa Barbara. A progressive journalist wrote that climate scientists like me were in it for the grant money—the exact argument the fossil fuel industry used to discredit climate scientists. My nervous system, responding to stress, exhaustion, and illness, went into fight or flight. I started waking up every night with a feeling like there was a bear outside my window.

Sonia and I talked daily, strategizing how to frame the bill in the press and keep up momentum. "Stay focused," she told me. "Just try to tune out the noise. None of this will matter in the long run if we get this bill passed." But it was hard to follow her advice. We are social creatures—our brains are wired for connection. Over the past few years, I had come to identify more and more with being a climate activist. So when a handful of people implied I didn't belong in the climate movement, it stung.

I wasn't the only one facing criticism. There were false claims that independent models—showing the bill would cut pollution by around 40 percent—were secretly funded by the fossil fuel industry. Some called for an unbiased government model, and when the Department of Energy released one confirming the previous estimates, they ignored it. I watched as groups twisted themselves into knots to avoid believing that the package, while flawed, would deliver progress. It was a hell of a lot better than nothing.

A few colleagues, watching these attacks, reached out to share their own experiences. Julian NoiseCat, having run the successful campaign to make Deb Haaland secretary of the interior, called to lend a supportive ear. He told me he'd been focusing on making a documentary and writing a book, and had mostly quit politics, disillusioned by all the drama. "It's strange," he said. "You and I are basically saying the same thing about the bill. But you're the one getting attacked." Other colleagues shared stories about the last days of the Paris climate negotiations in 2015. Although the agreement was now viewed as a great success, finalizing it led to infighting, as exhausted activists tore each other apart.

To be an activist, you had to be willing to fight. The trouble was, this energy could be turned inward toward one another rather than outward toward common opponents. In my first book, which studied how policy change happens, I'd coined a term: "the fog of enactment." The idea was that policymakers, advocates, and opponents often struggled to understand what a bill would do when it was passed. That was certainly playing out here. But I now saw another dynamic I'd missed: "the fight of enactment." When words were written on paper, people had to decide if it was good enough or not. And not everyone agreed. Maybe the final steps to pass a law were always a big fight and only viewed with rose-colored glasses in hindsight.

In an attempt to exploit this growing conflict among progressives, a fake grassroots organization appeared out of nowhere. Called "United for Clean Power," their website argued the bill should not pass because it was not good enough on environmental justice. It featured pictures of Representatives Alexandria Ocasio-Cortez, Ilhan Omar, Jamaal Bowman, and Rashida Tlaib—the most progressive House members. But it looked like it was written by someone's angry uncle who spent too much time on Facebook. Visually, it resembled a 1990s website made using MS Paint, with neon green Comic Sans font. Even its message was strange: "No Reconciliation Without Comprehensive Climate Change." For those keeping track, activists wanted to *stop* climate change, not ensure it was "comprehensive."

Reporting from independent journalist Emily Atkin showed that this shell organization was spending tens of thousands of dollars to try and derail the bill from passing. It was clear to me they were fossil fuel–funded, trying to trick progressives into believing the bill was bad by running ads on Facebook and Google and buying expensive spots in *Politico*. They also mass-texted constituents of progressive members of Congress, like Ocasio-Cortez. Some offices began hearing from these constituents who were being misled. In response, at least one environmental justice group called these members to reassure them they supported the bill and alert them to this fake group.

As the days passed, I began to reflect on where the sincere pushback came from. Before the bill had reappeared, several groups had campaigned for the president to declare a climate emergency. And they had come so close to seeing their goal achieved when Biden went to Massachusetts, only to watch him give a hollow speech. One of the biggest things a climate emergency would do was stop auctions for fossil fuels on public lands and waters. Not only had they lost that chance when the climate emergency was never declared, now the bill *required* auctions. It must have been a painful few weeks for these campaigners.

Supply-side groups focused on keeping fossil fuels in the ground, but that wasn't the bill's approach. It was a demand-side strategy, aiming to replace the machines that ran on fossil fuels to stop extraction in the medium term. But what if that never happened, and fossil fuel companies just found new ways to use their products in petrochemicals and plastics? What if communities just kept getting poisoned? What if funding for carbon sequestration just caused more pollution? These were fair concerns.

Ultimately, I had a different view. I believed that if we broke the tight correlation between economic growth and pollution, it would be easier to keep fossil fuels in the ground. If any American could get to work without burning oil, heat their home without gas, and cook a meal with clean electricity, it would be easier to take on the fossil fuel industry politically.

Most of all, I came to believe that attacking allies would not help our cause. We had to weave our strategies together, building a fabric that was stronger, even as strands pulled in opposite directions. A divided climate movement was the biggest gift we could give to the fossil fuel industry.

As outside conflict grew, it began to spill into Congress. Senator Bernie Sanders was angry about Manchin's requirement for fossil fuel leasing. Adrian met with his chief, to talk him through the bill's climate progress. Bernie had spoken to climate activist Bill McKibben, who told him he supported the bill. Varshini Prakash from Sunrise called him too. But Adrian still worried Sanders could turn against the package.

It was then that Adrian realized he had another problem. During the secret negotiations, he had used an old version of the energy tax title, before it had been scrubbed for the parliamentarian—and Manchin's team had spent ten days marking that version up. When the Finance Committee staff started working on the bill again, they quickly discovered the mistake. It caused chaos. Dozens and dozens of technical changes were missing. Now they had one week to fix them all while also preparing for the Byrd bath with the parliamentarian. It was going to be a very long week.

35

THE BYRD BATH

Just three days ahead of the Senate vote, Adrian woke in the middle of the night, suddenly alert. While deep in sleep, his mind had grasped something: a procedural error might give the Republicans a way to kill the entire bill.

That night before bed, he'd read a long, technical document: the Congressional Budget Office's updated estimates for the bill's cost. One provision—a new tax on pollution through the "Superfund"—was expected to generate around $11 billion. Yet CBO's report listed no government revenue from it. When Adrian read the table, he was confused, but didn't give it much attention. That is, until nearly 1:00 a.m. when his subconscious alerted him, waking him from sleep. "Oh my god," Adrian realized. "This provision is vulnerable to a privilege challenge."

The budget reconciliation process follows technical, arcane rules, and that CBO table suggested one of those rules might have been violated. If that was correct, the whole bill was dead. "We would have to start from scratch and pass it back out of the House," he thought. And there simply wasn't enough time left to do that. Adrian panicked. He grabbed his phone and fired off an email to a colleague, but she was asleep. For the next five hours, he tossed and turned in bed, alone in the world with this knowledge.

It would take mere hours for Republicans to come to the same conclusion, raising a challenge with the parliamentarian, claiming this procedural mistake meant the bill should lose its privileged status and require 60 votes to pass instead of a simple majority. Now Adrian would have to wait and see who the parliamentarian agreed with.

For the past several months, Adrian and his colleagues had been in constant meetings with the Senate parliamentarian, Elizabeth MacDonough. To prepare, they organized murder boards, with experts throwing hard questions at the presenting staff to sharpen their arguments. There were a handful of staff who had gone through reconciliation before and who proved particularly helpful at guiding the Democrats' strategy. Staffers had a habit of defending a provision on the grounds that it was "good policy," but arguments like that irritated MacDonough. Her job, she explained, was not to judge whether a policy was good or bad—it was to decide if it had budgetary impact.

The meeting tempo picked up in early July, when the bipartisan "Byrd bath" began in earnest. The process was named after Senator Robert Byrd, another powerful West Virginia Democrat. Under President Reagan, Republicans held just 53 seats in the Senate and were in the minority in the House. Yet they managed to pass several major reforms using reconciliation, including, most controversially, the 1981 Omnibus Budget Reconciliation Act that gutted Medicaid, Medicare, and Social Security benefits. The following year, reconciliation was used to change the number of seats on several regulatory commissions. Byrd objected to these policy changes being attached to budget bills just to circumvent the filibuster. In response, he established a series of rules for reconciliation in 1985. At their core, they required that every provision increase or decrease government revenue. The parliamentarian judged whether these directions were followed. Anything she ruled noncompliant became a "Byrd dropping."

But the rules also included other technicalities, like only allowing committees with instructions in the budget resolution to participate. If committees *without* instructions had parts of a reconciliation bill assigned to them, that could blow up the entire process. Think of a head chef standing at a pass, yelling out instructions to her various stations to cook the components of a dish, only to find her sous chef had thrown a sauce over the top that she didn't ask for. The dish would need to be cooked again from scratch—and it was the same with budget reconciliation if a committee added something that wasn't in the original instructions.

By the summer of 2022, Democratic staffers had met with the parliamentarian dozens of times about the bill. Even though the Republicans were often invited, they rarely participated. Perhaps they believed the

Democrats would never secure the votes, so why bother wasting their time. But when the package reappeared in late July, the Republicans showed up, throwing grenades.

Most meetings took place in person in the LBJ Room, just off the Senate floor. It was a small, ornate space with garnet walls, heavy wooden doors, and a ceiling of nineteenth-century frescos. The setting only heightened the drama of these face-offs. Staff from the majority and minority leaders' offices, the relevant committees, and the Budget Committee all crammed inside. Then MacDonough would enter alongside her deputies, take her place at the front, and preside over the meeting like a judge. Now in her mid-fifties, the previous year she'd been diagnosed with stage 3 breast cancer and had started treatment. With a bandana over her head, she looked frail. She was immunocompromised and asked all staff to wear masks at their meetings—even the Republicans complied. But despite her illness, MacDonough brought an intensity and eye for technical detail. She could remain focused in meetings that dragged on for hours.

On any given day, a Republican counsel presented an argument to the parliamentarian, explaining why a specific provision should be stripped for violating a Byrd rule. Democratic counsel or a member of Schumer's team then gave a rebuttal. Sometimes the arguments became so arcane that technical staffers were called on to elaborate. But none of this stumped the parliamentarian, who asked detailed questions, drilling into the claims. It went on like that: argument after argument, rebuttal after rebuttal, question after question. When the Republicans had exhausted all their attacks for the day, MacDonough stood to depart, her deputies trailing like ducklings behind her. Her final decisions would be delivered over email, in a day or two, as a simple judgment devoid of reasoning.

It soon became clear that the Democrats' extensive preparations were paying off. They were winning ruling after ruling. All that time they'd spent with the parliamentarian allowed them to preempt her concerns. "We thought that the Republicans were going to come at us hard and turn the bill into Swiss cheese," Adrian later reflected. "I am still in shock that we salvaged 95 percent or more of the package."

But then, mere days before the floor vote, Adrian woke in the middle of the night in terror. He realized the CBO had estimated zero dollars from the new tax because the revenue was earmarked to go back into the Superfund, to support pollution remediation, rather than the general fund, to balance the budget. The Senate Appropriations Committee, which oversees that Superfund, did not have any instructions in the budget resolution. Under the Byrd rules, they were not supposed to participate—they were like that sous chef, adding the wrong sauce to a dish. Still in bed, Adrian pulled up the relevant CBO document and saw this fact indicated with a small footnote buried on the very last page of the document. But even in small print, it was still there in plain sight: a way the Republicans could derail everything. If the parliamentarian agreed, the *entire bill* would lose its "privilege"—its ability to move quickly through the Senate with 50 votes. And this problem could not be overcome by an amendment. The House would need to start again, passing a new bill from scratch.

To avoid a catastrophic outcome like this, Schumer's team and other Senate staff had completed a "privilege scrub" back in November, combing through the 2,000 pages before Pelosi brought the bill to the House floor for a vote. They thought they'd scrubbed it clean. But they'd missed this change to the Superfund.

When everyone else was finally awake, Adrian talked to the budget staffers leading the reconciliation process, and they immediately understood the gravity of the situation. A few hours later, Republicans raised the challenge, and all the Democratic staffers panicked. Everyone except for Mike Jones, the Budget Committee's deputy staff director. A thoughtful man with horn-rimmed glasses, who paused to think before speaking, he'd served on the committee for over two decades. Mike's office was lined with stacks of paper laying out precedents from previous Byrd baths. This deep knowledge gave him a particularly good rapport with MacDonough. Now it was his job to figure out how Democrats would respond to this Republican challenge. The meeting with the parliamentarian was scheduled for the next evening.

At 7:00 p.m. that Friday night, Democratic staff from the Budget Committee and Schumer's team met with MacDonough to discuss the challenge. Mike argued the Republicans were wrong to raise it this late. The correct time to raise a privilege challenge was nine months earlier, when the House bill was first sent over to the Senate. He drew on a precedent

MacDonough herself had written in 2015. As she had put it: "In all but the most obvious and offensive cases, the process should move forward." He also argued that the Superfund fell within the Environment Committee's jurisdiction, which had reconciliation instructions. MacDonough listened intently and said she would email her decision, hopefully within the next few hours. There was nothing else to do but wait.

Adrian missed this meeting with the parliamentarian because he was busy dealing with another crisis. A week had passed since Schumer and Manchin's agreement was announced, yet Senator Kyrsten Sinema remained uncommitted to voting for the bill. On Wednesday, August 3, she gave Schumer an ultimatum: he must cut $14 billion in tax revenue and add $5 billion in "drought funding" for Arizona. Neither of these changes were acceptable to Manchin, who didn't want any senator messing with the delicate agreement. He had maxed out on spending and wanted the bill to raise enough revenue to reduce the national debt.

But the bigger problem with Sinema's demands lay elsewhere in the caucus, with the senators along the Colorado River. Her proposal wasn't just money for drought. It was a renegotiation of long-standing, very intense conflict between the upper and lower basin states of the Colorado River. "These people deeply distrust each other and have been fighting over this shit forever," Adrian thought. He knew he was walking into a hornet's nest.

With just two days before votes, Schumer called up Adrian and told him to fix it. And so he convened a complex negotiation between the eight senators from Arizona, California, Colorado, and New Mexico. It was another Western water fight, not unlike the Crooked River conflict Adrian had worked on a decade before. But this time he had hours to resolve the problem, not years. While he had played the role of patient peacemaker for Senator Merkley, working with Schumer had changed him. The situation demanded he make it clear he would not be messed with. "I had very little patience," Adrian reflected later. "I was basically pissing everybody off, but we didn't have time to deal with much bullshit. The stakes were very high."

On the surface, these Western senators faced a joint problem and could work together to convince Manchin they needed more funding.

The Western United States was in the midst of its worst drought in over a millennium. Senator Mark Kelly from Arizona and Senator Michael Bennet from Colorado both maintained a good rapport with Manchin, and each approached him. Kelly brought data, showing that Lake Mead and Lake Powell were each below 40 percent capacity after the long drought. He was able to get Manchin to agree to at least $1 billion. After that, Bennet persuaded Manchin to increase it to $4 billion.

But the bigger problem was how to divide the funding along the river. Bennet, representing the upper basin in Colorado, pushed for permanent reductions in water use and funding for public entities and Tribes. This was directly contrary to Sinema's demands for the lower basin in Arizona, which included temporary reductions and funding for irrigation districts. There was very little space for agreement, but Adrian realized that—apart from Sinema—none of these senators were willing to kill the climate bill. Not this late in the game. He had to call their bluff. He stayed up half the night on Thursday going back and forth with the senators' offices. They iterated on language until landing on something vague enough for everyone to project their own preferences onto. They could—and would—fight over the money again during implementation. But for now, they agreed.

The whole thing wrapped up on Friday, while his colleagues were busy meeting with the parliamentarian. Just after 9:00 p.m., she delivered her verdict. The Democrats were right: the bill could continue to the Senate floor the next day. Together, they'd performed two miracles that night. Adrian went home, kept working, then slept for a few hours. When he returned to the Capitol on Saturday, he brought his camping pad, in case he found time to nap on the floor. He knew it was going to be a long weekend.

36

27 HOURS

At 5:00 p.m. on Saturday, August 6, Adrian leaned against a red marble pillar at the back of the Senate chamber, his arms and legs crossed, bracing himself. Schumer had just raised the motion to proceed. At least 50 senators would now have to vote to consider the bill. And then they could debate it for the next 20 hours: a grueling vote-a-rama. During this time, an unlimited number of amendments could be raised. Adrian knew Republicans would take full advantage, trying to sow disarray with amendments designed to fuel attack ads during campaign season. Once they got through this gauntlet, the final vote would be called. It was Adrian's job to worry until that was over. Just how many hours it would take was anyone's guess.

Sonia also planned to stay up all night—on call for questions from senators. Republicans structured their amendments to maximize confusion, so Sonia followed closely, trying to understand the meaning and motivation behind each proposal. The process was extremely fragile, with no margin for error. Just because Schumer had the votes to proceed did not mean that the bill would pass.

Over the next 90 minutes, senators trickled into the chamber. Senator Patrick Leahy from Vermont, recovering from a broken hip, arrived in a wheelchair. He was greeted by Senator Whitehouse, who threw both hands in the air with excitement. A few minutes later, Manchin and Heinrich walked in together and both voted yes on the motion. Manchin and Warner talked for a few minutes before the West Virginian left. He wouldn't be back for several hours. Meanwhile, Sinema lingered, sitting down at a mahogany desk next to Republican Senator Todd Young from Indiana. They ended up talking for over 90 minutes. The remaining Republicans were slow to arrive. Two and a half hours passed before the last votes were registered, with Vice President Harris casting the tie-breaking vote.

By then, Adrian's worry had grown. There was a rumor going around that at least one senator had COVID and was still on the floor. Democrats needed to stay healthy. If anyone else got COVID—as Schumer and Manchin had within recent weeks—it could derail the entire process. Or worse: what if someone died? Adrian knew he wouldn't sleep at all.

Just before midnight, the vote-a-rama marathon finally started, but no one knew where the finish line was. It would only end after 20 hours or if Republicans were simply too tired to drag the process out any longer. When each amendment was introduced, there was one minute for debate from each party, followed by a ten-minute window for votes. In practice, it could take much longer. During votes for the American Rescue Plan in March 2021, one amendment was held open for almost twelve hours because of Manchin's opposition. Schumer's team worried that could happen again. But heading into the evening, Manchin assured Schumer's team he would stick with their agreement and vote down all Republican amendments. This was a huge relief for Adrian: he knew most of them were designed to appeal to Manchin directly.

Unity was the only way they'd get the bill done. They needed to vote down every amendment, because as soon as a single one passed, the deal would start falling apart. The Republicans, sensing weakness, would begin to exploit cracks. And once they got a taste of momentum, they would start having fun with it. The vote-a-rama would drag on and on, as they blew up the bill in a million tiny ways.

Early on, there were signs of fissures within the caucus. Once debate began, Senator Bernie Sanders stood to criticize the bill, calling it the "so-called Inflation Reduction Act." He introduced three amendments to strike all fossil fuel funding, create a Civilian Climate Corps, and reestablish the child tax credit. They were laudatory amendments that were not going to pass.

But the real problems emerged around 1:00 a.m. when Republican Senator Lindsey Graham from South Carolina offered an amendment on the Superfund, the same provision that had caused that last-minute drama with the parliamentarian. His choice of wording was particularly adept, calling it "a tax increase . . . on energy sources for Americans earning less than $400,000 per year." No Democrat wanted to vote against this provision—and vulnerable senators were particularly wary. But a split caucus so early in the process would be lethal.

Schumer, sensing they could lose the vote, held it open. He called several senators into his office and read them the riot act, saying if they didn't stick together on this, they could torch the deal. Did they want to be responsible for failing humanity? They came to a compromise, allowing five Democrats to vote for a second amendment on the same topic, which would still fail because it required 60 votes to pass. This unity set the tone. After that, everyone stuck together. That is, until the middle of the night, when Sinema started talking about tanking the bill.

Five days earlier, Anna Taylor had gone to the hospital. She was nine months pregnant and wasn't feeling well. The doctors diagnosed her with cholestasis, a liver condition, and decided to induce her. But labor was not easy, dragging on for days. Finally, her baby was born healthy, and they were both discharged. Shortly thereafter, Anna had to return because of postpartum complications. She was in the hospital during the vote-a-rama, when Senator Sinema started demanding changes.

Sinema, who did not get along with Manchin, was frustrated that more of her wishes weren't reflected in the final package. She'd worked for months on a plan to give new mothers paid leave. The whole thing was ready to go, but Manchin refused to expand government benefits in any way. After spending that night talking to Republicans, she decided instead that she was not happy about a new corporate minimum tax for firms with over $1 billion in annual profit. She also demanded to shield private equity firms from a hefty tax increase that Manchin wanted to put on them.

With her decade-plus of experience, Anna insisted on helping resolve Sinema's tricky demands during these middle-of-the-night negotiations. Gerry tried to tell her to rest and prioritize her health. But Anna wouldn't hear it. "If you stop including me, I will never forgive you, Gerry," she told him. "I will never speak to you again." From her hospital bed, Anna brokered the deal. The corporate minimum tax was tweaked, and the private equity tax increase was removed. That was enough to secure Sinema's vote.

It was now morning, and the bill still hadn't passed. After staying up all night, Senator Dianne Feinstein from California, who had just turned 89, was shivering. Sleep deprivation had compromised her body's thermoregulation. Other women in the caucus started taking care of her, bringing her blankets to warm up. Senator Amy Klobuchar of Minnesota found Gerry, updated him on how Feinstein was doing, told him to delay a vote, and ran off. Meanwhile, Schumer started yelling at senators to stop going to the bathroom. "You need to sit down and vote!" he said. Adrian anxiously watched it all unfold. What if someone ended up in the hospital?

Just before 3:00 p.m. on Sunday, the final amendment was considered. After that, the full bill would be voted on. Adrian and Gerry were standing at the back of the chamber when their chief came to find them. "Go sit down next to Schumer on the floor," he said. "This bill wouldn't have happened without you." Together, they walked down the center aisle and took the two open seats next to Schumer. Adrian had come to the Hill over a decade earlier to try to pass a climate bill. It had been three and a half years since he'd joined the majority leader's team to work on this package, then still a dream. He'd poured all his energy into this, sacrificing time with his young family. Now that it was finally happening, he felt like he was outside of his body, watching himself from above.

By the time the final votes started, more than 27 hours had elapsed. I thought the bill would pass in the middle of the night in California and I would miss the dramatic ending. But the delays allowed me to watch the final act on C-SPAN. I couldn't believe it when I turned on the livestream and saw my friend Adrian sitting right next to Senator Schumer. He looked exhausted.

"Madam President," Schumer began, as he addressed the entire chamber, "it's been a long, tough winding road. But at last, at last, we have arrived. I know it's been a long day and a long night, but we've gotten it done. Today, after more than a year of hard work, the Senate is making history. . . . To the tens of millions of young Americans who spent years marching, rallying, demanding that Congress act on climate change, this bill is for you." Schumer paused his speech, and the clerk called the final roll.

I watched as senators walked to the front of the chamber, one by one, recording their vote. Every single Republican voted against a livable future for their children, their nieces and nephews, their neighbors. And then they left the room. They had lost.

It had all been so close. If any Democrat fell sick or god forbid died, it would have fallen apart. I could see Senator Leahy being pushed around in his wheelchair. Senator Feinstein had nodded off at her desk.

When the final votes were recorded, the senators began to clap. Senator Brian Schatz ran straight up to Gerry, and they hugged. "We fucking did it!" the senator exclaimed. Schatz was from Hawai'i, and maybe it was the islands' politics, but he was a humble guy and treated Gerry like they were friends. They were also close in age: Schatz was 49, making him a veritable spring chicken in a caucus whose average age was 63. Climate was Schatz and Gerry's number-one issue, and they were both elated that they'd finally gotten it done.

Senator Tina Smith held her hands to her chest and wiped away a tear, as she headed to the back of the chamber to celebrate with more staff. Other senators patted each other on the back, hugged, shook hands. Everyone understood they had changed the course of history.

In their excitement, Democratic staffers flooded onto the floor, breaking Senate decorum. But no one corrected them or told them to leave. It was the appropriate response, given the gravity of the moment.

Schumer returned to his lectern for his closing remarks. An old man, with a neck hunched slightly forward, he prepared to speak about perhaps the biggest accomplishment of his long and productive life. Then Vice President Harris was handed a sheet of paper from the clerk. She read it aloud: "On this vote, the yeas are 50, the nays are 50. The Senate being equally divided, the vice president votes in the affirmative and the bill as amended is passed." The chamber erupted again.

When he began his remarks, Schumer fought back tears. "I've got to compose myself here," he said, before pausing briefly, the weight of the moment hitting him. "Every senator here knows an undeniable truth. We can never do what we do without our amazing, incredible staff. They work behind the scenes, they never fall under the spotlight, but . . . [they] worked overtime to help us in this historic endeavor, and you'll tell your grandchildren you were here, you were here."

He thanked his own staff by name, including Adrian, Gerry, Tim, and Anna, even shouting out her newborn daughter. "We heard her crying happily in the background as we were moving through all of this," Schumer said, smiling.

After he finished entering the names of all his staff into the record, Schumer walked away from the lectern and went off script. He turned to face the staffers now standing in the back of the chamber. "This bill is going to change America for decades. And you did it! Wherever you go, whatever you do, you should never forget how much you have helped make the world and the globe a better place. Never forget it." Watching at home, I felt Schumer was speaking to me. The tears came again.

That vote took 27 hours: the length of a long and painful labor. Adrian, Sonia, and I spent the time in a state of exhaustion and panic. But by the end, the largest climate investment in American history had finally passed the Senate. Together, we'd rewritten the future.

The next morning, Adrian woke up in his own bed. It was a Monday, but he wouldn't be going into the office. After months of delay, he was starting his paternity leave. The man who put on work clothes on Saturdays was finally lounging around in pajamas and hoodies again. While he'd delayed his paternity leave, their six-month-old had started to smile. And there he was on the couch with his dad, beaming up at his father.

ACT III

HOMEWARD

37

OF EARTH

After the Inflation Reduction Act passed the Senate, the House voted for it on Friday, August 12. It was now headed to President Biden's desk, and implementation would start immediately. Sonia decided it was time to call her boss.

When she reached him on Monday, she explained it was the honor of a lifetime to work at the White House, and she'd now accomplished what she came to do. "I no longer believe I'm the best person for the job," she said. A person with agency experience would be better suited to implement the law—someone who understood the federal bureaucracy like a pilot knew a flight path. Sonia wasn't ready for another year of endless calls with lawyers. She was ready to spend more time with her family.

But her boss wouldn't hear it—he thought Sonia understood the law's programs better than anyone. Behind the hundreds of pages of bill text were unwritten words about what each program was intended to do. "Please don't leave," he pleaded. "Replacing you will take time, if it's even possible." Sonia relented, agreeing to three more months. She would stay until the international climate conference in November.

The signing ceremony was scheduled for the next day in the White House State Dining Room. The space, often used for fancy dinners, was bright and ornately decorated, with high ceilings and historic portraits. That day, rows of chairs squeezed in dozens of politicians and reporters. Sonia was still in New Mexico and couldn't attend in person, so she joined on the official Zoom with the other remote staffers. Schumer gave opening remarks. "If there was a single word that could sum up this bill, it would be *persistence*," he said. "If you persist, you can make the world a better place—we persisted and persisted and never gave up. And today it pays off." He said his dad, who had died in the last year at 98, was looking down and smiling. Then Schumer turned to the president, who stood beside him, and shook his hand.

When Biden took the podium, he was in a good mood. Today was perhaps the most consequential of his presidency, and the crowd, which gave him a standing ovation, could feel history being made. "Chuck, you've been a good friend a long, long time," he said. In the audience he saw Senator Manchin had a black sling supporting his right arm—he'd had shoulder surgery during the recess. Looking straight at the senator, Biden said, "Joe, I never had a doubt," and sensing a warm audience, added to the joke. "Joe had an operation on his shoulder. I just want you to know it wasn't because of anything we did. He's in great shape." The crowd erupted in laughter.

Having dispensed with the pleasantries, Biden explained why this law was one of the most consequential in American history. "This bill is the biggest step forward on climate ever—ever—and it's going to allow, allow us to boldly take additional steps toward meeting all of my climate goals," he said. "It includes ensuring that we create clean energy opportunities in front-line and fence-line communities that have been smothered, smothered by the legacy of pollution, and fight environmental injustice that's been going on for so long."

When it came time to sign the bill, Sonia could not contain her excitement. Her colleagues spotted her on the screen, jumping up from her seat. Biden signed his name, then looked up at Manchin standing beside him, and handed him the pen.

By then, I'd recovered from COVID and returned to Canada to see my family. I hooked up my computer to the satellite internet and tuned in just in time to watch the scene unfold. I had to laugh when Manchin got the signing pen. In the end, he'd sold the bill on Fox News and helped push it over the finish line. Maybe he deserved the credit.

We were visiting my parents in the Northern Ontario woods, where I'd spent so much of my childhood, and I was trying to decompress. There was talk of me going into the White House to help implement the bill, but I wasn't sure that was what I needed. I had rashes on my hands, nerve pain in my right leg, and had stopped sleeping. Instead, I took a few weeks off from the firehose that is the internet. During that time, my mind wandered, and images from my childhood appeared like photographs: the tiny white flowers on our backyard hedge, the cedar trees that leaned out like elephant trunks over the lake, the paths I used to walk in the woods.

To help make sense of the conflict during bill passage, a friend recommended *Homage to Catalonia*. The book recounts George Orwell's experience in the Spanish Civil War, as he watched unity on the left devolve to infighting. Not to spoil the ending, but they lose to the fascists. I took the memoir with me on a solo overnight canoe trip in Algonquin Park and read it underneath the pine trees. My dad had grown up in this park, watching his parents cook meals and serve visitors at a wilderness lodge. Standing in the lake where he played as a child, my feet on its sandy bottom, I thought about my grandmother, who died more than a decade ago, and how far I'd traveled from this place I loved. Maybe my work had helped protect these lands and waters, so that my grandchildren could stand in this very spot long after I was gone.

I knew that while we'd won the battle, the war was far from over. I would spend the rest of my life fighting for more progress like this. I thought about how easily it could have all slipped away. It wasn't hard to imagine parallel universes where the bill failed to pass. In that late summer air, I could feel an oncoming chill. It was the same in the political world: the weather was shifting. Attention to climate change was waning. And with Twitter's sale to Elon Musk two months later, we would lose our most valuable campaigning platform. If we hadn't squeezed the bill through that summer, it would never have happened.

When I got back from that trip, I called a friend who had worked closely on the bill. He told me one utility company was already redoing its ten-year plan because of the Inflation Reduction Act, scrapping gas plants and building more clean power. Another was going to scale up its smart panel production, making it easier for people to electrify. And then there were the manufacturing investments, more each day: solar panel factories in Georgia, electric car battery facilities in North Carolina, wind energy manufacturing in Texas. All in mere days. "We will never do anything this consequential again in our lives," I told my friend.

Now that the bill was law, Sonia wanted to wrap her head around the scale of investment that would go out the door in the coming months and years. She took the top-line spending number—$370 billion—and divided it by the number of weeks until 2030. Her calculation suggested the federal government needed to spend around a billion dollars every week for the rest of the decade. But that was a low-end estimate. The funding could leverage up to ten times more in private capital. And the federal spending might be closer to $1 trillion, since many of the tax credits were uncapped. It wasn't clear that the government was up for the challenge of getting all this money out the door quickly. There were dozens of grant programs and tax credits to implement. And so many lawyers obsessed with following the process, rather than speed.

More than spending, Sonia began to fixate on time. She started a countdown clock on her desktop, with the number of weeks until 2030—385 when it began. In the coming years, that clock would become the calendar of her life. But for now, Sonia had just 21 more weeks in this job. She spent them trying to implement the three laws they'd managed to pass, which she affectionately called "Uncle Bill, Uncle Chip, and Uncle Ira."

Jahi Wise was also thinking about what he wanted to do next. He decided he should continue his mission to create a green bank and head to EPA to help design the process to implement that $20 billion. He wanted to make clean energy tangible for people across the country, helping them feel the positive effects in their communities.

Before wrapping up their time in the White House, Sonia and Jahi were promoted to commissioned officers. It gave them new privileges, including the ability to walk around the White House compound with greater freedom. They could now go over to the Navy Mess in the basement of the West Wing and order the special chocolate chip cookies everyone talked about.

Becoming a commissioned officer was also ceremonial, with a certificate from the president. The staffer involved with processing the paperwork asked Sonia what state she wanted listed on the certificate—like "Sonia Aggarwal of Ohio," for example.

Sonia didn't know how to respond. Sure, she was born in Ohio, but she'd gone to college in Pennsylvania, spent time in California, and now called New Mexico home. She didn't feel more attached to one of these places over the others. She loved the whole country, and more than that, the whole planet. "Could it say that I'm from the Earth? Like Sonia Aggarwal, of Earth?"

In his 20 years in the role, no one had ever asked him for anything like this. "Sure," he replied. "I'll see what I can do."

38

CELEBRATION

On Tuesday, September 13, Tim Ryder woke up with the song "Perfect Day" from the movie *Legally Blonde* running through his head. Rooting around in his closet, he grabbed his striped tie. The colors, cascading from blue to red, represented how much the planet had warmed since Tim was born. He'd bought it years before and wore it when the Senate passed the budget resolution that kicked off the long process for the climate bill. As the months passed, it became a lucky charm: anytime there was an important event or vote, he wore that tie. When high-level officials came in to meet with Schumer about the package, Tim was there in his striped tie. As they went through the Byrd bath with the parliamentarian, Tim wore the tie. And when the bill finally passed the Senate, Tim's tie could be seen on C-SPAN as he stood at the back of the chamber.

When he put it on that September morning, he knew it would probably be the last day he'd wear it. The president had invited a few thousand people who worked on the bill to celebrate its passage on the White House South Lawn. Many people were traveling from across the country to be there. Tim knew he would see hundreds of colleagues he'd barely, if ever, met in person. On his way out the door, he grabbed a black Sharpie on a whim. "I'll get a few folks to sign this tie," he thought. "It will remind me of all the people I love who worked with me on this historic bill."

It was a hot day, with no hint of the autumn to come. In his excitement, he forgot to apply sunscreen, and his cheeks turned redder and redder as the tie filled up with signatures. But Tim didn't care: he was having too much fun. He bumped into his Department of Energy colleagues, who excitedly signed it. Then he saw the modelers, who helped him estimate the bill's impact, and asked them to sign it too. Of course his closest collaborators, Sonia and Adrian, got prominent placement. And then I bumped into Tim and added my name.

As the temperature climbed, James Taylor took the stage and sang "America the Beautiful." After his performance, the Marine Band played "Hail to the Chief" as the president walked out of the White House. I watched as Vice President Harris, following behind him in a pale pink suit, took the lectern to begin the speeches. "Today, because of you—members of Congress, state and local officials, labor leaders, climate activists, and community organizers. And because of our president, Joe Biden—America is leading on climate, not just with declarations but with action," she said.

At the end of her remarks, she introduced a young woman named Lovette Jacobs, an IBEW electrician from Boston. Sporting a bright red shirt, she stood behind the presidential seal and began to speak, taken aback by the almost 5,000 people in the crowd. Harris stood beside her, encouraging her to take her time. That's when Lovette hit her stride. After becoming an electrical apprentice in 2017, she'd first met Biden while volunteering during his campaign in New Hampshire, waving a sign with other union workers. He came over to chat, and they struck up a conversation. The next time she saw him, she shared that she'd been laid off during the pandemic and was struggling to make ends meet. Biden showed her his trademark compassion, and in the next few years, Lovette watched as Biden supported unions with good-paying green jobs. "Mind you, the IBEW was one of his earliest supporters when he decided to run for president," she said. She turned toward him, and they beamed at each other, taking in the joy of that moment. "I'm proud to introduce the greatest champion of American workers. Ladies and gentlemen, the forty-sixth President of the United States, Joe Biden."

Biden hugged her, picked up a portable microphone, and stared out at the crowd through his signature Ray-Bans. He quickly listed some of the people who had made the law possible: Harris, Pelosi, Schumer, and Manchin. "Joe, thanks for sticking with what you said you'd do. I appreciate it," he said, pointing to the senator from West Virginia who was near the front. "This couldn't have happened without every single one of you—and that's in a literal sense in the Senate. Every single one was required because the other team didn't want to play . . . In the end, every single Republican voted against this historic law, so it fell to the Democrats to meet this moment and deliver for the American people, and that's exactly what we did."

"This year, the American people won, and the climate deniers lost," Biden said. He talked about how the law would make it cheaper for everyday Americans to buy a heat pump or an electric vehicle, how it would lower energy bills, and how it would spur clean manufacturing. "I want to take the most aggressive action ever, ever, ever to confront the climate crisis and increase our energy security—ever in the whole world," he said, the crowd erupting.

He also emphasized how the law would cut pollution for the most burdened communities, sharing his own story of growing up near oil refineries in Delaware. When his mom drove him to school, she would have to use the windshield wipers to remove the oil slick that had formed overnight. "That's why we had the highest cancer rates of any state in the nation for a long time," he reflected. It was this experience that made him want to deliver progress for these fence-line communities first. "All this progress was declared a failure . . . when we introduced it," he reminded the crowd. "We didn't give up, though. We kept at it. We had a vision, a plan, and we stuck to it. And the result is we're getting the job done for the American people, and we're just getting started. Together, we'll champion investments that lift all American working families and provide for the future of our kids."

As I stood there watching, I thought about all the people who had made this moment possible. The young organizers in Sunrise, like Varshini Prakash, Nikayla Jefferson, and Kidus Girma, who poured all their energy into building a movement. The older activists, like Bill McKibben and Jamie Henn, who had been sounding the alarm for decades. I thought about politicians like Representative Alexandria Ocasio-Cortez, who had launched the Green New Deal just four years earlier. And others like Senator Ed Markey and Governor Jay Inslee, who had worked toward a federal climate bill for many years. I thought about the staffers in Congress, who'd crafted policy for decades, waiting for the chance to finally pass it into law. I thought about folks in the White House, who'd aged visibly as they pushed this bill over the finish line. I thought about the strategists in think tanks and nonprofits, like Christy Goldfuss; the modelers, like Jesse Jenkins; and the other academics, like me. I thought about the unions and companies who finally got behind clean energy. The journalists telling the stories, the funders supporting the activists, the dreamers and schemers. We'd needed everyone to get this done.

Crammed into the crowd with Sonia beside me, I felt lost in the sea of people. But as soon as the speeches finished, the crowd thinned, and there among the trees were all my friends. Sam Ricketts and Julian Noise-Cat and Varshini Prakash and Keane Bhatt and Adrian Deveny and Tim Ryder and hundreds of others. We hugged, laughed, and took photo after photo. For the next eight hours, we went from party to party, collecting more friends at each stop.

As I headed out for the night, Julian NoiseCat pulled me aside. We hadn't worked together for years now, as he'd been making a documentary and writing a book about Indigenous life in North America. But we'd stayed friends and were thrilled to celebrate together that night. "It was wonderful to spend this chapter, whatever this was, working with you," he said. "You should write a book about it."

The next morning, Tim woke up with a new idea: "I should get Schumer, Pelosi, and Biden to sign the tie too!" His boss was easy. Schumer had a deep affection for Tim, so he just walked into his office and he signed it. The senator found the project endearing. A few weeks later, they met with Speaker Pelosi's team, and she added her signature, too. President Biden proved the hardest to track down. Tim tried reaching out to a friend who worked in the White House, but he laughed, brushing him off.

In the end, he had to wait a full year. On the one-year anniversary of the law, Tim was staffing Schumer at a White House meeting with the president, and he wore his tie. As they waited just outside the Green Room, Tim reminded the senator that he needed his help to get Biden to sign it. "One hundred percent—let's do this!" Schumer said. But when the time came, the Secret Service informed Schumer that the president wanted to meet with him alone. Staff would have to wait outside. Schumer was about to walk through the door, when he suddenly turned around and exclaimed, "Tim, your tie!" Awkwardly, with the Secret Service watching the strange scene, Tim yanked it off his neck and handed it to Schumer. When he returned, the president's signature was on it. Now Tim could finally retire that tie.

39

FAREWELL

As Sonia's last day approached, she found herself having second thoughts. Was she making the right decision? A job in the White House allowed a person to have more impact in a day than they might otherwise have in a lifetime. Maybe it was never the right time to leave a job like that.

For her send-off, her colleagues organized a series of Zoom celebrations. She joined one call, unsure of what to expect, and found a dozen people there to recount stories. Sonia was mortified: it was hard for her to be the center of attention.

Her boss recalled a time during the campaign when she'd calculated how much pollution could be cut by Biden's latest climate plan. He presented Sonia's work to then-Vice President Biden. With his tendency to get into the weeds, Biden pored over the document carefully. "Have you done the math?" he asked. "Are you sure this is possible, that we can cut pollution this fast?" "Absolutely," her boss replied. "I've got a rocket scientist doing the math!" Sonia did have a degree in astronomy and physics and had spent time observing faraway planets. She wasn't quite a rocket scientist, but it was close enough. There were few people who could count tons of pollution as accurately as Sonia.

Another colleague told a story from the transition, when Sonia gave a briefing. Her newborn son was lying on her chest, visible on the screen. Despite Sonia's impassioned delivery, he slept through the entire call. Maybe he'd heard her powerful voice thousands of times in the womb. Now that he was a toddler, he became a regular in the background of work calls. Her colleagues laughed about how they often heard him singing his made-up songs.

Sonia was known around the office for her personal motto: "Get tons." She'd made it her life's mission to cut as many tons of carbon pollution as possible. This perspective allowed her to bring a rare creativity and determination to her job. Everyone agreed she'd made the legislation she worked on better. She never stopped believing they could pass more ambitious laws—and she'd turned that belief into reality through hard work. In countless ways, for generations to come, Sonia had changed the world.

When it was her turn to speak, Sonia struggled to keep it together. Her voice quavering, she told her colleagues how grateful she was to work alongside them. Ever the analyst, she spoke in numbers. Together, they'd set the country on track to cut a fifth of its annual pollution. A billion tons by 2030, to be exact. "Of course, we all know how much work it took to get to that pen stroke and how much work lies in front of us," Sonia said. "We need to do this, and we need to go even faster because there is no bigger prize than keeping our kids healthy, preserving this incredible human species, and making the lives of people and animals all across the world a little easier for thousands of years to come."

By November, she'd taken her last trip to DC as a White House employee and handed in her badge.

In the coming months, many other people who worked on the climate law would leave their jobs. By the time Sonia left, Gina McCarthy was already gone, and John Podesta joined the White House to implement the Inflation Reduction Act. In November, Jahi Wise left to work on the green bank at the EPA. Then in January, Anna Taylor decided she wanted to spend more time with her daughter and left Schumer's team after more than a decade. In February, Brian Deese announced he would leave the White House, and by the end of the month, Ron Klain was gone too.

When Gerry Petrella decided to leave in April, he'd spent over a dozen years by Schumer's side. His going-away party was jam-packed, with tons of senators and even Speaker Pelosi. His mom, who was there that day, couldn't understand it. He was just her kid from Long Island, but somehow Pelosi—whom she loved—was celebrating him. At the party, Schumer gave him one of the pens he'd received from President Biden at the Inflation Reduction Act signing ceremony. Gerry had finally caught his white whale.

Outside of government, climate activists also left their posts. That winter, I decided to leave Evergreen, and soon I would be gone from Rewiring America too. Within the year, Sam Ricketts left his leadership role at Evergreen. Christy Goldfuss moved from the Center for American Progress to help run the Natural Resources Defense Council. At Sunrise, Varshini Prakash announced she would be leaving in the spring of 2023, after seven years as executive director. And environmental justice advocate Cinthia Moore, who had championed electric school buses in Nevada, successfully ran for office. Now she was fighting for her community from the State Assembly.

There was far less turnover within Congress. Nancy Pelosi did step down as speaker of the house, but Schumer remained majority leader and President Biden, who was in his early eighties, insisted that he was running for another term. Always a person of the highest integrity, it was Senator Tina Smith from Minnesota, whom I'd worked so closely with, that announced she was retiring in her late sixties to make space for the next generation of leaders. Replacing a climate champion like Smith would not be easy. But I was happy for her. She was going to spend more time with her grandchildren.

Adrian brought his trademark overthinking to quitting his job. It never felt like the right time, but he knew he couldn't keep working eighty-hour weeks and be the kind of father and partner he wanted to be. It would be another year before he found himself surrounded by colleagues, being toasted for his years of work. On that day, he felt the same ambivalence that Sonia had: the job he was leaving had allowed him to change history. He likely wouldn't get another chance to make that big of an impact for the rest of his life.

Adrian's send-off took place in person on a Friday afternoon in early February 2024. His friends and colleagues gathered in the hearing room for the Environment Committee. They mingled in front of the dais where the senators sat, sipping cans of seltzer and beer. Adrian had been in this room dozens of times, with its mid-century wood walls, brass sconces, and well-worn carpet. The ceiling was painted with a series of white and blue squares that held astrological signs and symbols: he'd long heard rumors that the architect was a fan of horoscopes. Adrian gazed up at the ceiling, wondering if he'd ever look at these paintings again.

For the occasion, he wore a light blue tie and his navy suit. The pockets were now loose, stretched out from years of holding two phones: one for his personal life and a second for work. In a sea of dark oxfords, he stood out with his tan monk strap shoes, buckled at the side. Adrian never followed *all* the rules.

His wife and kids were invited, but she was running late. She picked up their daughter early from school, raced across the city, and still got stuck in traffic. When they finally parked, she ran inside, dragging their five-year-old behind her. In the end, the effort was worth it. People kept thanking her, as if she had passed the law herself. And in some sense, it was true: Adrian could not have done it without her support. She only wished their parents could have been there. Adrian's in-laws had gone from wondering why he was always on the phone late at night to telling everyone they knew that their son-in-law had changed the world.

Eventually, Senator Schumer came to address the assembled group. Photographers moved between staffers to record the event, their large flashes going off periodically. Schumer understood the power of an image, always making sure cameras were present if he was giving a speech. By then, he'd served in elected office for almost 50 years. He'd passed dozens of laws: first as an assemblymember in New York, then in the House of Representatives, and finally in the Senate. He'd been a senator for more than a quarter of a century. Hundreds of people had worked for him over the years.

"Most of my staff, when they leave the Hill, they can say they've changed the world and made it a better place," Schumer began. "But I think few have really made it more of a better place than Adrian. He was able to actually do more about climate change than probably just about anybody else. . . . There would not be an IRA without Adrian. It's very rare you can say that about a single staffer, as good as people are, that it wouldn't have happened without them. But that is true with him. He could go sit on his couch for the rest of his life—which he won't do. But he could do nothing else. And he still would've accomplished more than 99.9 percent of what other people accomplish. It's an amazingly gratifying feeling for him, I'm sure. And for his family," he said, looking at Adrian's wife and daughter.

"He encouraged me at moments—me, the perpetual optimist who thought we'd never get it done! He just kept going and going. And when he hit a wall, he found a new way around the wall. There were a lot of walls. I could name some of the specific senators who were walls . . . but I won't on this happy occasion. He found a way to get the job done.

"So it is a sad day in the sense that one of the best staffers that any senator could ever have—and that is not hyperbole—is leaving. It's a happy day for him. He's a real family man. The only thing that tortured him more than not getting this done was being away from his family. And he had to be away so often to do this with two beautiful children.

"Adrian, you're a treasure. You're a treasure as a human being, as a person, as a friend, as a staffer, as a loyal comrade in the fight to preserve our globe," Schumer said, in closing. "It wouldn't have happened without you."

Adrian's five-year-old daughter beamed up at her dad, marveling at all the people who had come to celebrate him, unsure exactly what for. She knew he had done something big, and was now leaving his job, but the details remained blurry. That afternoon, when Adrian walked out of the building with his young family, he left one of his phones behind.

After the law passed, Adrian gave his friend Hasan Nazar a framed copy of the provision they'd worked on—the one that invested in clean energy manufacturing across the country, shaving thousands off the cost of an electric vehicle by subsidizing battery production. Hasan proudly displayed it in his office.

At the time, Hasan was still working at Tesla. But that spring, Elon Musk decided they no longer needed a public policy team, eliminating it wholesale. Despite delivering billions of dollars of value to the company, Hasan got caught up in this mass exodus. That same month, Musk, then the richest man in the world, started a political action committee to support Donald Trump's latest presidential campaign. By July, he had endorsed Trump.

40

HAPPY BIRTHDAY

In the summer of 2023, the Inflation Reduction Act turned one year old, and my daughters turned two. Their birthdays would always be a week apart. This time, we didn't have COVID, and I wasn't working endless hours. To celebrate, we had friends over to our home, where we had just added solar and heat pumps using the tax credits I'd helped pass. It was now fully electrified, and I'd paid my last dollar to a gas company. That felt great.

For the occasion, I wore a long lavender dress and baked a sesame cake, elaborately decorated with peaches from our garden. My daughters sat on the living room floor with my husband as I lit the candles and we all sang. They marveled at the tiny fire, trying to blow it out with little success.

It felt painful to think back two years, to that day in the hospital when I was rushed into surgery for an emergency delivery. We didn't know if the babies would survive or suffer profound difficulties all their lives. After they were discharged from the NICU, my daughters continued to be monitored. Just before turning two, they were assessed—and the results were extremely positive. Their language was advanced. Both girls were so thrilled when they communicated an idea or got an adult to do something they wanted. They loved looking at picture books, with paintings by Mayumi Oda, photographs of animals, and images of gardens.

Despite the difficult pregnancy and months in the hospital, they were both now healthy. When a mother gives birth, she often experiences a powerful flush of oxytocin, bonding her to her newborn. Given I'd had a c-section at just 29 weeks, I never experienced this hormonal surge. And for the first week of their young lives, I couldn't even touch or hold them. These strange and scary circumstances made it harder for me to feel that instantaneous connection. Instead, it took time to bond with them, and I found my love expanding as weeks, months, and years passed.

It was the same with the Inflation Reduction Act. When it was first signed into law, I was too exhausted to celebrate. But in the months that followed, a deep sense of pride grew inside me. When a coal plant was scheduled to close early, I knew I'd helped make that possible. When a heat pump factory was announced, I knew I'd supported its funding. When states made plans to cut pollution, I could see them using the dollars I'd helped secure. And it was the same for so many others: thousands of people were proud of helping to pass this law.

In that first year of the law's life, these same people worked to implement it. Implementation required a deeper level of attention, and no one person could track more than a few programs in detail. The environmental movement began to branch out, like a river forking in many directions. As the months passed, I talked to my activist friends less and less. I decided to focus on working with environmental justice groups to try to maximize state electrification funding flowing out of the Department of Energy. We aimed to deliver as much as possible to lower-income Americans and minimize any funding for gas furnaces. My friends at the Natural Resources Defense Council put together a broad coalition to fight for a hydrogen tax credit that was as green as possible, and I helped with that when I could. Ricketts pored his attention into the $5 billion in state grants we'd helped secure. Even deep-red places like Utah, Oklahoma, and Texas applied, with about $33 billion in total requests. When the clean electricity program had fallen apart, we were right to ask for so much funding for states. They were ready to lead.

At the Environmental Protection Agency, Jahi spent his time trying to ensure that the green bank would help people across the country. When Jahi was just a few weeks into his new job, Senator Bernie Sanders started demanding to know what was happening with the $7 billion in solar funding he'd made happen. Sanders had dreamed for years about a program to help lower income folks put solar on their roofs, just like he had done on his own home. But he worried that the EPA wasn't doing what he intended under the law. He was so upset, he'd put a hold on confirming a nominee for the agency until they could sort this out.

Together, Jahi and the EPA administrator walked over to Senator Sanders's office on the Hill. They found him in a bad mood. "I understand we're here so you can tell me how you're planning to implement this solar program," Sanders said gruffly, looking straight at them. His boss turned to Jahi, saying, "Yes. Please explain to the senator our plans." After two years in the White House, Jahi knew how to think on his feet. Plus, he'd built a deep understanding of the gaps in solar deployment. He quickly laid out their plans to launch the program. Sanders, nodding along, softened as Jahi spoke. When the meeting wrapped, he turned to Jahi and said, "You seem very serious about this." "Of course, senator," Jahi replied. "This is very serious. I am looking forward to working closely with you." Shortly after, Sanders voted for the EPA nominee. As the months passed, Jahi worked hard alongside dozens of colleagues to try to get the money awarded and out the door. But moving billions of dollars through an agency built to regulate, not spend, was challenging—and slow.

The Treasury Department was similarly working around the clock to launch the bill's more than 40 energy tax credits. By the end of 2023, that manufacturing policy, which Hasan worked on with Adrian, was projected to cost almost $73 billion—making it the largest in the package. Plus, it was leveraging hundreds of billions more in private funding. Hasan thought back to how little attention the provision had received when the law passed, and had to laugh. He'd known it would be transformative, but it became much bigger than even he'd imagined. Crucially, around 80 percent of this investment was in Republican states. Adrian and Hasan's idea to shift the politics was starting to pay off.

As the law rippled out across the country, many environmental groups who had opposed it during passage became vocal supporters. For one thing, their worst fears of a fossil fuel leasing boom did not come true: when some auctions were held, there were almost no bidders. A year into implementation, I met with one activist who had changed her opinion about the law and, by extension, me. We found much more common ground than I expected. "There's a perfectionism to the climate movement," she told me when we met in person. "I'm a perfectionist myself—and it's something I'm working on." I found her honest and insightful. Activists know how little time there is to act on the climate crisis, and the dire consequences if we fail. But we can sometimes struggle to distinguish between progress and perfection. And we can glorify process over outcomes.

Not everyone felt that the law's widespread success was a good thing. By the time the first anniversary rolled around, Manchin had soured on the package, claiming the Biden administration had implemented it as a "radical climate agenda." Perhaps the senator had been caught in the fog of enactment, unable to anticipate the law's profound changes to the energy system. But his views wouldn't be relevant much longer. With West Virginia's popular governor challenging him, Manchin decided to retire rather than lose his seat.

Just after the bill's first anniversary, in the fall of 2023, I returned to DC for a work trip. It was unseasonably warm, at over 100°F. Climate change had not stopped just because we passed one law. In some ways, it was even more painful to watch the impacts now that we'd finally acted. When Canada's boreal forest burned that spring, blanketing the East Coast in thick smog, I felt heartbroken. The lungs of our Earth were burning, and now they were burning our lungs. These ecosystems that had protected us for so long were harming us—because we had harmed them. It wasn't supposed to be this way. But even on these hard days, I knew there would be fewer tragedies like this because of our work. I would never have been able to say that if I'd focused only on my own small carbon footprint. This progress was only possible by working through our democracy, joining with thousands of others to create a big carbon wave.

On that fall trip, I met with staff in the Biden administration and Congress, sharing advice on how to implement the law. Sonia happened to be in town, so we met up with Adrian. It was the first time the three of us had ever hung out in person. Adrian, still working for Schumer, wore a suit and melted in the heat. He talked about quitting his job soon. Sonia was now running a big climate organization, helping to drive strategy and funding for the movement. Having these two people in my life, who saw the world like I did and worked to cut pollution every day, was a blessing beyond measure. And if Sonia and Adrian existed, there must be tens of thousands more like them, in every corner of the planet, linked in a vast web of action.

Now, renewables were surging. Solar, batteries, and wind would end up supplying 96 percent of new electricity capacity in 2024. With only one new gas plant in the entire country, clean energy was basically the only thing being built. And supply chain problems meant that even if a utility wanted more gas, they would have to wait upward of seven years just to get a turbine.

The law was catalyzing real projects in every corner of the country. Thanks to Brian Deese's efforts, the number of fast EV-charging stations had tripled since Biden took office: by 2025, there were 12,000 across the country. No one would have trouble finding a charger like he did years earlier.

Adrian found it particularly gratifying to see so much manufacturing growth in Georgia. If Warnock and Ossoff hadn't won their Senate seats in the runoffs, the climate bill would never have passed.

Outside Cleveland, Ohio, near where Sonia grew up, $129 million in funding was now flowing to several huge solar and battery projects to replace a coal plant, which had operated since 1888. For 136 years, kids like her had played outside, breathing polluted air from that plant. But now, because we'd secured state and local funding after the clean electricity program died, it would finally shut down.

For Ricketts, his deepest joy was found in the bill itself. Even two years after it became law, he found himself rereading the text, marveling at all the threads he could connect to earlier policy proposals. When he first started working in Congress in 2009, he remembered being summoned one afternoon to the speaker's office, just off the Capitol Rotunda. When he walked inside and noticed the elegant chandeliers and historic portraits, he knew he was in a special place. He was there to present Pelosi's top advisors with a list of economic recovery ideas from the House environment caucus. All the most seasoned Hill staffers were there that day, true veterans with the gray hairs to prove it. And there was Ricketts, just 23 years old, with the audacity to tell them he had the plan. Now he was 37, and so many of those ideas had become law. He knew his mom would have been proud.

But the summer of 2024 was hardly a time for celebration. When Biden debated Trump in June, he did a terrible job. The polling was awful, and I could see he wasn't going to win the election. He probably should never have run for a second term and stuck to his word that he planned to be a one-term "transitional" president, passing the torch to a younger candidate. Still, I felt bad for Biden, who'd been the most progressive president in my lifetime and had delivered on climate. But most Americans didn't see him the way I did. They saw him as too old for another term. When I went on a radio show for the climate law's second anniversary, praising Biden fell flat with listeners. The November election was still months away, but I had a sinking feeling it wouldn't go well. And I worried what that would do to our law, still only a toddler, let alone our democracy.

But even if the law was repealed, I knew many of its impacts could never be reversed. There were already millions more electric vehicles on the road, solar panels on roofs, and heat pumps in people's homes. The cost reductions in all these technologies would stick around too.

There was always going to be more work to do, regardless of who won the election. We needed to eliminate all carbon pollution by 2050, a deadline smack at the end of my working life. I knew I would spend my career cranking away on this one problem. Once, we were young activists who inherited the climate crisis. But we were not so young anymore. It was now our responsibility to solve it. Working together, we can break the chain and refuse to pass this problem along to the next generation.

In the end, having children did not narrow my focus like I feared. In an unexpected way, it expanded it, connecting me more deeply to all life. When I heard stories of children harmed by fossil fuel pollution or killed in war, I felt their pain in my bones. Sometimes, in my daughters' tiny faces with their big eyes, I could glimpse other baby animals: kittens and birds and even fish. So much of our genetic code is shared across species. This scientific fact felt less abstract now. All of life is related. The work was to protect this beautiful, interwoven world not just for *my* children but for all living beings.

As they grew, I also learned about myself. I could see their budding concern for the life around them. If they dropped a cracker on the ground, they would say it was "for the birds." Their sensitivity to touch and sound and light mirrored my own, giving me a new understanding of the body I'd been born into and now passed along. It connected me to my parents and grandparents and ancestors I had never met, tracing back until humans were other species. It connected me more clearly to the air and the ocean. That ocean we came from, in the salty brined belly of our mothers. That ocean we will return to after we die. That ocean that dances with the air and absorbs all that extra carbon we've put into the atmosphere. That ocean is us. Somehow I knew that more deeply than before.

Years later, when I looked at my children, now walking, now talking, now marveling at the world, I knew, beyond a doubt, I'd made our planet a better place for them to grow up in. There is no better feeling than this.

EPILOGUE

Just a few weeks after President Biden signed the climate law, back in August 2022, I got a call from Ike Irby. A climate scientist by training, Ike had worked for Vice President Kamala Harris for five years. We met in 2019, when Harris was still a senator and was running in the presidential primary. He was smart and nerdy, and we hit it off immediately. Once Harris became VP, I heard from Ike occasionally, between his endless trips crisscrossing the country in Air Force Two.

"We've been thinking about doing an event in California with Vice President Harris, talking about what the administration has accomplished on climate," he told me that afternoon, as I sat in my car parked on the side of the road.

"That's a great idea!" I responded, rattling off a bunch of possible participants.

"I don't think you understand," Ike said, cutting me off. "I'm proposing that you interview her live, for your podcast."

That October, I found myself walking to the old Fort Mason army post on the San Francisco Bay. It was a beautiful fall day, the sky free of the fog that could hover for weeks. After clearing security, my cohost Katharine Wilkinson and I stood backstage, reviewing our cue cards. We'd been making our show for two years but had never done a live episode, let alone interviewed the most powerful woman in America. Peering out at the crowd, the auditorium was packed. Normally, I was confident in situations like these—my years in high school drama class had trained me well. But watching the vice president walk toward us, surrounded by security carrying guns, made me nervous.

The mayor of San Francisco opened the event, and before we knew it, Katharine and I were seated across from Harris, in her sleek, dark brown suit. The energy in the room felt electric, more like a rock concert than a lecture. I'd heard the vice president's residence had recently added a heat pump and was excited to ask about it. "A heat pump is an efficient electric machine that can both heat and cool your home—no fossil fuels required," I said, and applause filled the room. Who knew that simply defining a technology would elicit such enthusiasm: clearly, we were playing to an easy audience. I rambled for a while, then tried to turn toward a question. "I could go on and on about heat pumps . . . But this isn't about me, this is about you. So I would love to know . . ."

"What do I think about heat pumps?" the vice president interjected, and we all laughed.

"Well, obviously, you must love them," I continued. "But what are some parts of these amazing new laws that you are most excited about?

As a senator, Harris had championed electric school buses, alongside grassroots environmental justice activists. Then as VP, she worked to secure $5 billion for the effort in the bipartisan infrastructure law. "I love electric school buses," she said that afternoon. "I really do. And we're manufacturing them in our country. I've been to the manufacturing plants. I've been on these electric school buses."

Every day, 25 million kids rode a school bus and breathed in their toxic diesel fumes. Because of Harris and environmental justice advocates like Cinthia Moore, almost 200,000 kids were now riding a clean bus to school. There was still a lot more progress to be made, but it was a start.

I left the event impressed. The vice president was fun. She brought enthusiasm and intelligence to the climate problem and talked eloquently about the pollution burden poorer communities carried. The laughter we all shared in the auditorium spilled out into the street, as we spent the rest of the day celebrating with friends from across the climate and environmental justice movements.

Two years later, that joyous afternoon felt hard to conjure. Like millions of other Americans, I spent the summer of 2024 in a state of increasing anxiety. After bombing the June debate, I felt certain Biden couldn't win the election. That feeling only grew after the terrible assassination attempt against Trump. But weeks went by, and Biden remained in the race. Instead of focusing their attention on beating Trump, Democrats descended into infighting, arguing about whether or not Biden should drop out. The party was in complete disarray, while democracy and the fate of our planet were both on the line.

On Sunday, July 21, I decided to take a break from the news, and drove a few miles out of town with my daughters to the Santa Ynez hills. We were going raspberry picking. Just days away from their third birthday, they loved running between the rows and putting more berries in their mouths than their baskets. I was standing out in the field when I heard that Biden had dropped out and would support Harris. Minutes later, Ike called. "The next few days are critical to solidify support behind Harris," he told me. "I'll do everything I can to support her," I said, becoming emotional. Maybe now we'd have a chance to protect our democracy and lock in all the climate progress we'd fought so hard for.

For the next 100 days, I threw myself into supporting the campaign: organizing fundraisers, writing op-eds, going on TV, and doing endless press interviews. It was the first time since the climate law passed that the climate and environmental justice movements felt unified again. We all knew that a second Trump term would be disastrous for the planet. The Heritage Foundation, a conservative think tank long involved in climate denial, had written a 900-page roadmap for a future Trump administration. Their plan was to dismantle the climate law, federal climate science, and clean air protections. Journalist Maxine Joselow had also reported that Trump told fossil fuel executives to donate $1 billion to his campaign and he would do anything they wanted. Environmental activists understood that for the planet, for people, and for our democracy, the stakes of this election could not be higher.

On election day, Adrian was visiting us in Santa Barbara to give a talk about his time working for Schumer. By then, he was an independent consultant, organizing a vast climate policy network to feed ideas into a possible Harris administration. At lunch, we exchanged forecasts for the night ahead. "I think she's going to win," I told him, despite growing more worried over the past two weeks. "I hope you're right, but I'm not feeling that optimistic," Adrian said. "I think it's a coin flip. Trump seems to have a lot of momentum in rural parts of the country."

As we sat down in my living room to watch the returns roll in, Adrian warned that I might find him annoying. He would be toggling between the phone, computer, and TV screen, searching for the earliest data in South Florida or rural Georgia. It didn't bother me: this was exactly how my husband did election night. And within 30 minutes of the first polls closing, neither of them were feeling good. "If these trends continue," Adrian said, "she's going to lose." The same feeling of dread we'd all experienced in 2016 returned, though this time with less shock. We went to bed early, knowing Trump was going to be president again.

Two days later, Adrian and I were sitting onstage, talking to a packed room of students. Most were working toward their master's in environmental science and management, and the audience was grim. One of the first things we did was provide context. Around the world, four in five incumbent governments lost reelection that year—despite usually having the advantage. The long tail of COVID, with its lockdowns and economic disruptions, had led people to blame their governments. Adrian shared stories of friends who had long voted for Democrats who turned to Trump after their businesses went under during the pandemic. "They didn't have a chance yet to vote on that issue," he explained. "So this was the election where they expressed their critique of the system for the first time."

And while Harris had run for president for 107 days, Trump had been doing it for a decade—2,615 days to be exact. After he mailed checks with his name on them during the pandemic and staged a photo-op pretending to work at McDonald's, many Americans believed Trump would help with their financial problems.

As Adrian and I talked to the students that evening, there were just 74 days until inauguration. The Biden administration would need to race to get money from the climate law out the door. Senator Bernie Sanders had been right in the end: we should have focused more on programs that could be scaled up quickly, giving Americans tangible benefits fast. Instead, the election had been lost, and billions still sat in government coffers.

Adrian, Sonia, Ricketts, and I spent those waning days of the Biden administration supporting government staffers trying to protect funds, making them harder to repeal. Of course, that assumed the incoming Trump administration would follow the rule of law.

And although Elon Musk seemed less and less like a man who'd made a fortune in the clean technologies of the future, I reasoned he still had an incentive to protect these climate policies. After spending $290 million on the election, he was the largest individual donor, and he had the president's ear—at least for now. Maybe he would use his access to advocate for electric vehicles, solar, and batteries.

On Inauguration Day, I called Ricketts instead of watching the event. After years of talking daily, we now went weeks without hearing from each other. Ricketts had started a consulting company focused on helping states make climate progress, while I had taken a break from working formally with climate groups and was spending time just being a professor and a mother. We both shared our worries about the destruction to come. "If they want to, they can mess with almost everything," Ricketts told me. "But you know, we've got to pace ourselves. Trump is kind of like the sun: you can't stare directly at him. You've just got to stay focused on making your corner of the world a little better." We reminded ourselves there were still things Trump couldn't do. The law had been in place for two and a half years by the time he took office. He couldn't uninstall a heat pump purring away in someone's garage or rip down solar panels from someone's roof. After we hung up, I felt better.

The first few months of the Trump administration proved worse than we expected. The president started coloring outside the lines, simply ignoring the rule of law. Without any justification, the administration began harassing organizations that had received funding from the climate bill, attempting to claw back federal dollars that were already out the door. Lawyers mobilized to defend these groups, and sometimes won in court. But it was painful to see so much progress dismantled.

In April 2025, as Trump became obsessed with tariffs, his Commerce Department slapped absurd 3,500 percent duties on solar panels from Southeast Asia. This marked the end of the pause Sonia had helped create three years earlier. To make matters worse, that summer, a trade court found the Biden pause was illegal, retroactively imposing up to $54 billion in solar tariffs. But this did not erase the importance of Sonia's work. Domestic solar manufacturing had ramped up 500 percent since June 2022. And it wasn't just solar panels being made in America: electric vehicle, battery, heat pump, and critical minerals manufacturing had all surged. Since the law passed, $170 billion of private capital had been invested in clean tech manufacturing, creating 165,000 new jobs. With nearly three-quarters in Republican districts, most were also better-paying than other jobs in the area. Republicans representing these districts now had a stake in the clean economy future.

But there were limits to what Sonia, Ricketts, Adrian, or I could do to defend these investments. Republicans in Congress did not care what we thought. Now the law we'd helped create was out in the world and out of our hands. It was almost like having children: sure, you created them, but your control shrinks with time. It was up to companies and Republicans to form their own carbon wave. This wasn't easy: the main trade association for the clean energy industry still had fossil fuel interests embedded within it. And while clean energy companies had grown, their influence over the Republican Party was much weaker than the fossil fuel industry.

That spring, as the first draft of the House Republicans' reconciliation bill began to trickle out of committees, we initially felt hopeful. I was particularly shocked that the home electrification rebates I'd poured so much time into were left untouched. Other programs, like the state grants Ricketts and I had helped design, were repealed, but almost all of the $5 billion had already been spent, including significant grants to states that Republicans governed.

As the bill wound its way through the House, the cuts became deeper. Before passage, the Committee on Rules made changes in the middle of the night, terminating the tax credits for wind, solar, and batteries within two and a half years while slapping on impossible red tape. When it passed the House, I felt defeated. "Don't worry—this is the worst it will get," one advocate told me. "The Senate will improve the bill."

Initially, this seemed likely, with a handful of Republican senators voicing strong concerns. Elon Musk also became a vocal critic, calling the bill a "disgusting abomination" before his falling out with Trump. Over the next month, Senate committees made tweaks, but whenever bill text was released, it was worse than we expected. Still, during the middle of the night vote-a-rama, Republican Senator Lisa Murkowski from Alaska pushed for key changes to protect solar and wind projects and fully preserve battery funding. The latter fact made me hopeful: while solar costs had fallen dramatically, batteries still had a ways to go. With a few more years of federal support, batteries plus solar would be locked in as the cheapest energy option.

After Trump signed the bill into law on July 4, things took another turn for the worse. A series of executive orders undercut the Republican senators' attempts to preserve solar and wind funding. Weeks later, the Interior Department imposed additional red tape for clean energy on public lands. Then the admin attempted to claw back more funding that was already allocated, including Sanders's $7 billion for solar—although several states and groups sued, arguing this was unlawful. The combined effect was to make the clean energy transition slower and more unequal. Rich people will always be able to invest in cleaner, healthier technologies for their homes. But lower-income Americans will struggle to keep up. These terrible changes, amid the rapid dismantling of democracy, led me to avoid reading the news.

I knew the climate movement, myself included, had made some mistakes. We had fixated too much on process at the expense of speed. And our perfectionism left us vulnerable to distraction. While we were right to want benefits to flow to organized labor, disadvantaged communities, and everyday Americans, the Biden Administration took too long to implement too much of the climate law. The assumption that some people made, that there would be four more years to get things right, turned out to be wrong. President Trump's disinterest in following the rule of law meant that even programs whose funds had been allocated were vulnerable to being clawed back. Even if Sanders' solar program wasn't frozen, the fact was it was too slow to get solar projects built. Tax credits could have simply been made available to lower-income Americans through direct pay. They would have had three years to put solar panels on their roofs, just like wealthier Americans were doing all that time. To deliver equity I knew that speed needed to be prioritized next time.

Still, when it came to climate progress, I began to wonder if the second Trump administration would prove a period of treading water rather than drifting backward. His policies would drive up clean energy costs, but they wouldn't create a fossil fuel renaissance. When the tax credits to help everyday people buy an electric vehicle, solar panels, and batteries began to expire at the end of 2025, millions of Americans rushed to use them. In the third quarter of 2025, electric vehicle purchases were a record 437,000 and residential battery sales were up 70 percent year over year. Particularly since the EV tax credits were going to be harder to use in future years, bringing forward this demand was a good thing. Plus, automakers dropped prices once the tax credits were gone—in some cases by more than the tax credits were worth in the first place. The AI boom, combined with transportation and building electrification, created a hunger for more electricity, and solar and batteries would continue to be built.

While the clean energy industry was poised for a hard few years, the fossil fuel industry was too. With falling oil prices and increased costs from tariffs, wells across the country were idling by summer. As one oil executive put it in August 2025, "We've gone from drill, baby, drill to wait, baby, wait." Plus, the supply chain for gas power plants was broken, making it nearly impossible for new infrastructure to be built before the 2030s. If President Trump drove the economy into a recession, pollution would fall from behavior change. This wasn't a lasting solution, but it could limit Trump's damage and buy us some time. The fact was fossil fuels were not thriving under Trump's second term.

Reading an author's back catalogue can often feel like experiencing the same story again and again with different covers. When I set out to write *The Carbon Wave*, I thought it would be a significant departure from my first book, *Short Circuiting Policy*. In that case, I examined what happens when clean energy laws are rolled back. After the Inflation Reduction Act passed in 2022, perhaps feeling a vested interest in the law's success, I was sure it would stick around. But when Trump took office, it felt like I was reliving the same story: watching climate policy take two steps forward only to take another one back.

I decided to reread my first book, looking to my past research for clues for the future. Back in 2020, I wrote: "If we want policy to lock-in, and drive long term change, we have to destroy the political status quo. This is not easy to do." These words reminded me to take a longer view. The fossil fuel industry is perhaps the most powerful in human history. We were never going to defeat them with one law. These setbacks are heartbreaking, especially given how little time we have left to cut carbon pollution. But we have to remember there are battles within a war. The fossil fuel industry is fighting so hard because we're winning that war.

During the 1970s energy crisis, the first modern wind turbines and solar projects were built because of President Jimmy Carter's policies. When Ronald Reagan was elected president, he rolled these policies back, causing bankruptcies across the nascent clean energy industry. Another energy crisis in the early 1990s brought a new tax credit for wind energy. Over the next three decades, the credit would lapse and be renewed several times. This is the story of clean energy: it oscillates up and down, much as we might hope it would grow in a straight line. We have to ride those downward dips, gathering our energy again to make the upward waves we need.

We don't yet know if we will be able to rip out all of the world's fossil fuel infrastructure by 2050. But I know that it's up to all of us—including you, the reader of this book—to try to make that happen. If you've made it to the end of this story, you very likely want to do something about the climate crisis. In reading these pages, I hope you've found some guidance on focusing your energy to make your carbon wave as big as possible. It could start with getting rid of the fossil fuel infrastructure in your own home, electrifying it with heat pumps, an induction stove and electric vehicle or bike, maybe putting solar on your roof and a battery in your garage. Or if you don't own a home, you could help a parent or friend electrify theirs. Your carbon wave could grow to working on your community, building clean energy, and electrifying public buildings. You could even find yourself changing local, state, or federal laws. While I don't know you, I can tell that you're not that different from Adrian, Sonia, Ricketts, and the dozens of other people in this book. By joining with others, you are capable of making big changes. This is the power of democracy.

I think back to 2018, when this story began. It was in the resistance to Trump's first term that Democrats found their focus on climate. During the 2018 midterms, Representative Alexandria Ocasio-Cortez won her seat running on a Green New Deal. Mere days later, she joined Varshini Prakash and other Sunrise activists for a sit-in at Speaker Nancy Pelosi's office. A few days after that, Adrian Deveny was finally called back for an interview with Senator Chuck Schumer. Months passed, then Governor Jay Inslee joined the presidential race, giving Sam Ricketts an excuse to write several hundred pages of policy, fleshing out the Green New Deal. President Biden won that election, and Sonia Aggarwal got a chance to serve in the White House. And, improbably, I started working with all of them to try to change the course of history. Looking back, I can see how the resistance to Trump set up a series of dominos, which began to fall, one after the other, until the largest climate law in American history passed four years later.

And now, we'll have to do it all over again. Next time, I hope you'll join our carbon wave.

December 2025

Santa Barbara, California

ACKNOWLEDGMENTS

On April 1, 2021, I started a new document. At the time, I was pregnant and working on what I hoped would become a climate law. I wanted to write down what was happening as a record to my future self. In the coming years, my children were born, the climate law passed, and my notes grew to over a hundred pages. I'd started to think there was a book to be written about what I'd lived through.

Writing groups kept me accountable and at my desk throughout this project. Thanks to my friends Justin de Benedictis-Kessner, Rebecca Donner, Dena Enos, Jess Green, Kate Marvel, AJ Rice, Kim Vaz-Deville, and Katharine Wilkinson for keeping me inspired and writing alongside me.

This book was written in dozens of locations over many years, but I am particularly grateful to the Harvard Radcliffe Institute for giving me a fellowship with an office for a year; to Sarah Anderson, Sage Davis, Charlyna DeFeyter, and Steve Gaines at the Bren School for an office with an ocean view for two years; and to Tzeporah Berman, Christiana Figueres, and Rebecca Solnit for a magical few days at Hollyhock on Cortes Island, Canada, where I wrote the opening scene of this book. Funding for this project also came from the 11th Hour Project and the Anton Vonk Chair. To my patron saints Shanti Kleiman and Diane Boss, I am forever grateful.

The Cottage Hospital NICU was full of magical people who kept my children alive and healthy. I could not be more indebted to Gail Moran, Kim Bowman, Ann Jenkins-Diebold, Jill Effling, Dr. Catherine Ekwa-Ekoko, Dr. Alex Soffici, and Dr. Melissa Drake.

UC Santa Barbara has proven to be an ideal place for engaged scholarship on climate, especially thanks to the leadership of Dean Charlie Hale. I'm grateful to my academic colleagues and mentors, including

Amit Ahuja, Michaël Aklin, Bruce Bimber, John Bowers, Kate Bruhn, Alison Brysk, Mark Buntaine, Liz Carlisle, Kim Cobb, Jeff Colgan, Ranjit Deshmukh, Joan Dudney, Ryan Enos, Alfredo Gonzalez, Jake Grumbach, Jacob Hacker, Jen Hadden, Tom Hale, Kathy Harrison, Katharine Hayhoe, Robert Heilmayr, Dan Kammen, Kent Jennings, Bob Keohane, Tae-Yeoun Keum, David Konisky, Paasha Mahdavi, Ed Maibach, Cesar Martinez-Alvarez, Eric Masanet, Dan Masterson, Jamila Michener, Julia Morse, Neil Narang, Will Nomikos, Andrew Norris, Ryoko Oono, Chris Parker, Paul Pierson, Barry Rabe, Timmons Roberts, Michael Ross, Maya Sen, Satyajit Singh, Eric Smith, Kathy Thelen, Dustin Tingley, David Victor, John Woolley, and Grace Wu.

My friends listened to me complain about book writing for years. Thank you to Kathy Burba, Louis Cassano, Bernice Chau, Laura Chirot, Marina Cortbawi, Amy Gordon, Tim Hade, Alex Hertel-Fernandez, Sandy Hudson, Romi Kadri, Rachael King, Ivo Lamers, Kelsey Perry, Olivia Quinn, Elana Resnick, Olivia Scobie, Dhara Solanki, and Beth Sternlieb. An extra thanks to my decades-long friends Danielle Westbrook and Tamara Stein for providing feedback on early drafts. And to Gail Moran for reading the manuscript, taking care of our babies, and providing endless support. Somehow I ended up with the coolest mentors, including Bill McKibben, David Pellow, Kim Stanley Robinson, and Rebecca Solnit. They are as smart, kind, and generous as you imagine.

As I hope the book makes clear, there were thousands of people involved in passing the climate law. A few of them became my close friends in the process. Thank you especially to Jade Begay, Sam Krasnow, Arjun Krishnaswami, and Camila Thorndike. I'm also grateful to the journalists—some of whom became like friends—who kept the bill as a front-page story for years. I reference much of their work in the endnotes.

Dozens of people agreed to be interviewed for this book, sharing their personal and professional experience. My heartfelt thanks to everyone for their willingness to share. I hope they feel their stories were captured in these pages.

This book would not have been possible without the friendship and trust of Sonia Aggarwal, Adrian Deveny, and Sam Ricketts. Each of them put up with hours-long interviews, endless follow-up questions, and multiple rounds of fact-checking. Their willingness to let me tell their stories is a gift I can never repay. I am also grateful to their children and especially their partners, Jordan MacHardy, Shahrzad Rezvani, and Alison Ricketts. Their work would not have happened without their spouses' sacrifices and support.

This book would not exist if it wasn't for my agent, Rachel Vogel, and my editor, Beth Clevenger. Both of them believed in this project when it felt like no one else did. Their endless edits, ideas, and encouragement made the book stronger. Thanks also to everyone at the MIT Press who worked on this project, including Judy Feldmann, Rachel Fudge, Debora Kuan, Summer Porter, and Karl Spurzem. Thanks also to Mark Drajem and Laura Fraser for feedback on the manuscript. For a full two years, as I wrote and rewrote the draft, Maddy Ranalli worked alongside me: fact-checking, researching, copyediting, and boosting my morale. Together, we edited until Google Docs informed us we had reached the comment limit for the file—who knew that existed. It was a pleasure and privilege to work with you, Maddy. Lucas Boyd similarly spent years supporting this project, with editing and encouragement. Without him, every sports metaphor in this book would be wrong.

And, finally, to my family. To my grandmother, who read 90 books a year and would have loved this one, and to my grandfather for four decades of laughter and stories. To my parents for their decades of love and support, and now for loving my children. To my sisters for friendship and long-distance calls. To my nephew and new niece for their brilliance. To my cat, Delilah, for joy. And to my husband, Matto Mildenberger: thank you for getting up every night for the past four years to reassure the girls, and for being a true partner in all areas of our life. I could not have done any of this without you. And, finally, to our daughters, who are smart, funny, kind, and full of light. You bring me joy beyond words.

ENDNOTES

This book relied on my firsthand experience as well as dozens of interviews. While I was working on the climate bill, I took notes, including verbatim quotations. In other cases, quotations were recollected by interview subjects from memory or notes, and some have been lightly condensed or edited for clarity. In a handful of cases, names were changed to protect an individual's privacy. In other instances, specific people were not named, for the sake of momentum in the storytelling—I tried to give them credit in the below endnotes. The draft also went through extensive fact-checking, including with the key interview subjects. As usual: all mistakes are my own.

For those interested in using this book in a classroom setting, and seeking an academic, analytical article to accompany the text, I recommend "The Politics of American Clean Energy and Climate Policy: Why the Inflation Reduction Act Passed," co-authored with Olivia B. Quinn, published in *Climatic Change*.

As I hope the book makes clear, this carbon wave took thousands of people working over many decades to pass the largest climate law in American history. Apart from those already named in the book, I would like to acknowledge the contributions of: John Agan, Margie Alt, Karen Anderson, Bobby Andres, Carol Andress, Sean Babington, Manish Bapna, Jeremiah Baumann, Trent Bauserman, Eddie Bautista, Lindsey Baxter Griffith, Ben Beachy, Matt Bearzotti, Jade Begay, Deborah Behles, Holly Bender, Jonathan Black, Renae Black, Christine Blackburn, Mikayla Bodey, Robert Bonnie, Anaïs Borja, Heather Boushey, Sage Briscoe, Roxanne Brown, Mike Brune, Robert Bullard, Holly Burke, Dallas Burtraw, Brennen Cain, Sam Calisch, Steve Capanna, Greg Carlock, Tom Carlson, Chris Carnevale, Mike Carr, Alison Cassady, Mark Chambers, Sara Chieffo, Heather Clark, Steve Clemmer, John Coequyt, Heather Coleman, Catherine Coleman Flowers, Jamie Consuegra, Kristina Costa, Cammie Croft, Ramón Cruz, Mike Darner, Bill Dauster, Christopher Davis, Kyle Davis, Shelby Davis-Prettiman, Kenneth DeGraff, Jamie DeMarco, Abbie Dillen, Paul Donohoo-Vallett, Greg Dotson, Patrick Drupp, Dan Dudis, Kerry Duggan, Rick Duke, Kristin Eberhard, Becca Ellison, Jess Ennis, Dan Esposito, Mike Evans, Todd Fernandez, Adam Fischer, Mike Fishman, Jane Flegal, Cara Fogler, Sam Fowler, Josh Freed, Zach Friedman, Carla Frisch, Amy Fuerstenau, Raúl García, Rose Garr, Rebecca Gibbons, Laura Haynes Gillam, Wes Gobar, Michael Goo, Rachael Grace, Rob Gramlich, Saul

Griffith, Rhiana Gunn-Wright, Warren Gunnels, Josh Gutmaker, Caitlin Haberman, Tim Hade, Katherine Hamilton, Jeff Hammarlund, Seth Hanlon, Aliya Haq, Marty Hayden, David Hayes, Bracken Hendricks, Trevor Higgins, Ethan Hinch, Mary Anne Hitt, Sam Holman, Zealan Hoover, Abby Hopper, Holmes Hummel, Dani Hupper, Bart Johnsen-Harris, Ayana Elizabeth Johnson, Dana Johnson, Mike Jones, Brian Kahn, Gene Karpinski, Noah Kaufman, Cathleen Kelly, Nat Keohane, Rick Kessler, Sam Krasnow, Mark Kresowik, Arjun Krishnaswami, Brendan Larkin, John Larsen, William Lawrence, Sarah Lazarovic, Saul Levin, Jamal Lewis, Alice Lin, Bill Lipton, Lori Lodes, Jeff Lomonaco, Mike Lynch, Dustin Maghamfar, Arjun Makhijani, Rev. Michael Malcom, Silvio Marcacci, Brad Markell, Alex McDonough, Sean McElwee, Yvonne McIntyre, Duncan Meisel, Russell Mendell, Maurice Mitchell, Harold Mitchell Jr., Lena Moffitt, Marcela Mulholland, Josh Nassar, Spencer Nelson, Bruce Nilles, Ali Nouri, Mike O'Boyle, Collin O'Mara, Robbie Orsvis, C. J. Osman, Arnab Pal, Daniel Palken, Steve Pantano, Bill Parsons, Rachel Patterson, Elizabeth Perera, Alisa Petersen, Amol Phadke, Melinda Pierce, Zach Pierce, Alex Piper, Whitney Potter Schwartz, Rich Powell, Tim Profeta, Olivia Quinn, Jamal Raad, Andrew Reagan, Kevin Rennert, Mary Frances Repko, John Richards, Michele Roberts, John Romankiewicz, David Rosner, Costa Samaras, Dan Schory, Ben Schreiber, Reed Schuler, Quentin Scott, Curtis Seymour, David Shadburn, Jigar Shah, Tarek Shah, Deirdre Shelly, Peggy Shepard, Aru Shiney-Ajay, Joe Shultz, Tony Sirna, Tiernan Sittenfeld, Karen Skelton, Lora Snyder, Jean Su, Narayan Subramanian, Aaron Suntag, Meghan Taira, Louisa Terrell, Nicole Teutschel, Maggie Thomas, Katie Thomas Carol, Camila Thorndike, Mike Tidwell, Todd Tucker, Ana Unruh Cohen, Candace Vahlsing, Uday Varadarajan, Hannah Vogel, Beth Vrabel, Jason Walsh, Sam Walsh, Lindsey Walter, Becca Ward, Evan Weber, Marc Weiss, Kathleen Welch, Lance West, Mike Williams, Ryan Wiser, Andrew Wishnia, Aimee Witteman, Beverly Wright, Pete Wyckoff, Elizabeth Yeampierre, Rev. Lennox Yearwood, Anna Yelverton, Miya Yoshitani, Ali Zaidi, Avi Zevin, Ethan Zindler, Adam Zipkin, Adam Zurofsky, and thousands of others. I wish I could have told more of their stories in these pages. My sincere apologies to people I've missed.

The following works aided in my construction of this story: Nick Estes's *Our History Is the Future*, Franklin Foer's *The Last Politician*, Ryan Grim's *The Squad*, Joe Manchin's *Dead Center*, and the documentary *An Inconvenient Truth*. I also relied heavily on secondary reporting from journalists, as cited below.

EPIGRAPH

Page vii **Something is wrong**: Mary Oliver, *Upstream: Selected Essays* (Penguin, 2016), x.

INTRODUCTION

Page 1 **a line in a platform**: Jay Inslee for President, "100% Clean Energy for America Plan," May 3, 2019.

Page 1 **grew into a policy paper**: Leah Stokes, Sam Ricketts, and Olivia Quinn, *A Roadmap to Clean Electricity by 2035* (Evergreen Collaborative & Data for Progress, February 2021).

Page 1 **President Joe Biden to commit**: Katie Glueck and Lisa Friedman, "Biden Announces $2 Trillion Climate Plan," *New York Times*, July 14, 2020.

Page 2 **made a fortune in coal**: Geoff Dembicki, "Joe Manchin Makes $500K a Year from One of the Dirtiest Coal Plants in West Virginia," *VICE News*, July 28, 2021.

Page 2 **on the chopping block**: Coral Davenport, "Key to Biden's Climate Agenda Likely to Be Cut Because of Manchin Opposition," *New York Times*, October 15, 2021.

Page 5 **9 percent in an instant**: Peter Gorrie, "Toronto Hits Energy Target," *Toronto Star*, March 30, 2008.

Page 6 **spent upward of $100 million**: Gregory Solman, "BP: Coloring Public Opinion?," *Adweek*, January 14, 2008.

Page 6 **Academic articles**: Seth Wynes and Kimberly Nicholas, "The Climate Mitigation Gap: Education and Government Recommendations Miss the Most Effective Individual Actions," *Environmental Research Letters* 2, no. 12 (2017): 074024.

Page 6 **16 tons of carbon pollution every year**: "Calculate Your Carbon Footprint," The Nature Conservancy, nature.org/en-us/get-involved/how-to-help/carbon-footprint-calculator/.

Page 9 **hundreds of advocates**: This network included individuals from Bluegreen Alliance, Center for American Progress, Chesapeake Climate Action Network, Clean Energy Buyers Association, Earthjustice, Environmental Defense Fund, Energy Innovation, Just Transition Alliance, League of Conservation Voters, Mom's Clean Air Force, National Resources Defense Council, Partnership for Southern Equity, People's Justice Council, Resources for the Future, Sunrise Movement, WE ACT for Environmental Justice, and many other groups.

Page 9 **made up the majority**: US Energy Information Administration, "More Than 60% of Electric Generating Capacity Installed in 2018 Was Fueled by Natural Gas," March 11, 2019.

Page 9 **96 percent of new additions**: Dan McCarthy, "Chart: 96 Percent of New US Power Capacity Was Carbon-Free in 2024," *Canary Media*, January 10, 2025.

Page 9 **$170 billion of private investment**: Clean Economy Tracker, "Data Collection Methodology," 2025, cleaneconomytracker.org/.

Page 9 **outsold gas furnaces**: Elena Shao, "As Heat Pumps Go Mainstream, a Big Question: Can They Handle Real Cold?," *New York Times*, February 22, 2023.

Page 10 **public wants climate action**: Jennifer R. Marlon et al., "Change in US State-Level Public Opinion About Climate Change: 2008–2020," *Environmental Research*

Letters 17 (2022): 124046; Aaron M. McCright, "The Social Bases of Climate Change Knowledge, Concern, and Policy Support in the U.S. General Public," *Hofstra Law Review* 37, no. 4 (2009).

CHAPTER 1: THE GREEN NEW DEAL

Page 15 **found its center**: Nick Estes, *Our History Is the Future: Standing Rock Versus the Dakota Access Pipeline, and the Long Tradition of Indigenous Resistance* (Haymarket, 2024).

Page 16 **emissions to fall**: IPCC, *Global Warming of 1.5°C Special Report* (2018).

Page 18 **were already cleaner**: Mike Scott, "Yes, Electric Cars Are Cleaner, Even When the Power Comes from Coal," *Forbes*, March 30, 2020.

Page 19 **with a scientist**: Amol Phadke at Lawrence Berkeley National Laboratory and UC Berkeley.

Page 19 **made up around half**: Carrie Schoeneberger, Jennifer B. Dunn, and Eric Masanet, "Technical, Environmental, and Economic Analysis Comparing Low-Carbon Industrial Process Heat Options in U.S. Poly(vinyl chloride) and Ethylene Manufacturing Facilities," *Environmental Science & Technology* 58, no. 11 (2024).

Page 20 **promoted climate denial**: Emily Williams, Sydney A. Bartone, Emma K. Swanson, and Leah C. Stokes, "The American Electric Utility Industry's Role in Promoting Climate Denial, Doubt, and Delay," *Environmental Research Letters* 17, no. 9 (2022).

CHAPTER 2: RIGHT PLACE, RIGHT TIME

Page 21 **one in 20 bills**: "Statistics and Historical Comparison," Govtrack.us, 2025.

Page 22 **an all-time high**: Catherine Bruns, Ami Nacu-Schmidt, Max Boykoff, and Meaghan Daly, "US Newspaper Coverage of Climate Change or Global Warming, 2000–2025—August 2025".

CHAPTER 3: CAMPAIGN SEASON

Page 27 **it was the top issue**: League of Conservation Voters, "New 2020 Poll: Climate and Clean Energy—Winning Issues in Primary & General," June 25, 2019.

Page 28 **skeptical that he would prioritize**: Benjy Sarlin, "Jay Inslee Tops Greenpeace Climate Grades, Joe Biden Gets a D-," *NBC News*, May 29, 2019.

Page 30 **Inslee's campaign staff**: After working for Jay Inslee's 2020 presidential campaign, Evergreen co-founder Maggie Thomas was hired by the Warren campaign.

Page 34 **others on the campaign**: Bracken Hendricks was a key author of these policy plans.

Page 35 **pollution fell by just 7 percent**: Pierre Friedlingstein et al., "Global Carbon Budget 2020," *Earth System Science Data* 2, no. 4 (2020): 3269–3340.

Page 35 **limit warming to 1.5°C**: IPCC, *Global Warming of 1.5°C Special Report.*

Page 36 **environmental justice leaders**: Leaders providing input to the Biden campaign included Harold Mitchell Jr., Cecilia Martinez, Catherine Coleman Flowers, and Michele Roberts.

Page 37 **a detailed roadmap**: The report was co-authored with Olivia Quinn, Narayan Subramanian, and Bracken Hendricks and supported by Jamal Raad, Julian NoiseCat, and Sean McElwee.

CHAPTER 4: RECONCILIATION

Page 40 **committee staff**: Ana Unruh Cohen was the staff director for the House Natural Resources Committee and spearheaded the effort.

Page 41 **economists had pressed Congress**: Mark K. Matthews, "Inside Economists' Decades Long Slog on Carbon Taxes," *E&E News*, June 25, 2019.

Page 41 **worked better on paper**: Matto Mildenberger and Leah Stokes, "The Trouble with Carbon Pricing," *Boston Review*, September 24, 2020.

Page 44 **polling showed**: Nathaniel Rakich, "Final Forecast: Democrats Have a 3-in-4 Chance of Flipping the Senate," *FiveThirtyEight*, November 3, 2020.

CHAPTER 5: ELECTION DAY

Page 46 **report she'd published**: *The 2035 Report* (University of California, Berkeley's Goldman School of Public Policy and GridLab, June 2020).

Page 49 **textbook called *Sustainable Energy***: Jefferson W. Tester, Elisabeth M. Drake, Michael J. Driscoll, Michael W. Golay, and William A. Peters, *Sustainable Energy: Choosing Among Options* (MIT Press, 2005).

Page 53 **front page of newspapers**: Somini Sengupta, "At Least 100 Dead in India Terror Attacks," *New York Times*, November 26, 2008.

CHAPTER 6: CONCEPTION

Page 57 **more than one in three Americans**: *Progress Toward 100% Clean Energy in Cities & States Across the US* (UCLA Luskin Center for Innovation, November 2019).

Page 57 **more likely to live**: NAACP, Indigenous Environmental Network, and Little Village Environmental Justice Organization, *Coal Blooded: Putting Profits Before People* (2012).

Page 59 **published a report**: Saul Griffith and Sam Calisch, *No Place Like Home* (Rewiring America, 2020).

CHAPTER 7: MOMENTUM

Page 62 **it happened every day**: "The Financial Crisis: This Day—September, 2008," *CNBC*, September 28, 2009.

Page 63 **helped bring back 115,000 jobs**: Stephen Koff, "Economic Adviser Brian Deese Says Auto Industry Has Rebounded with 115,000 Jobs," *PolitiFact*, June 9, 2011.

Page 64 **around a third**: InfluenceMap, "Who Owns the World's Fossil Fuels?," December 2018.

Page 64 **some climate groups pushed back**: Zack Colman, "Climate Purity Tests for Biden Nominees Split Enviros," *Politico*, December 3, 2020.

Page 64 **half degree centigrade**: Matt McGrath, "Climate Change: 'Monumental' Deal to Cut HFCs, Fastest Growing Greenhouse Gases," *BBC*, October 15, 2016.

Page 67 **accomplished almost nothing**: Katie Rogers, "How 'Infrastructure Week' Became a Long-Running Joke," *New York Times*, May 19, 2022.

Page 69 **helping design the administration's commitment**: Special Presidential Envoy for Climate John Kerry's team worked on this effort, particularly Rick Duke.

Page 69 **more than 6,600 MMTs annually**: Environmental Protection Agency, "Inventory of U.S. Greenhouse Gas Emissions and Sinks," 2022.

Page 71 **her bosses**: Sonia's other boss was Ali Zaidi, who served as the White House deputy climate advisor. He was also present for many of the scenes in this book featuring Gina and Sonia and played a key role in all climate policy.

CHAPTER 8: THE BUDGET CHAIRMAN

Page 73 **top issue for young people**: Alec Tyson, Brian Kennedy, and Cary Funk, "Gen Z, Millennials Stand Out for Climate Change Activism, Social Media Engagement with Issue," Pew Research Center, May 26, 2021.

Page 74 **the Finance Committee approved**: Under a power-sharing agreement, Democrats chaired the Senate committees and could move tied legislation out of committee using a discharge petition; Senate Committee on Finance, "Senate Finance Committee Advances Clean Energy for America Act," May 26, 2021.

Page 75 **installed solar on his home**: K. Kaufmann, "Politicians Talk a Green Game, but Do They Have Solar on Their Home Rooftops?," *pv magazine*, October 19, 2020.

Page 75 **cosponsoring bipartisan legislation**: Senator Bernie Sanders, "Sanders, Boozman Introduce 10 Million Solar Roofs Bill," press release, June 6, 2011.

Page 76 **Warner criticized it**: Alexander Bolton, "Sanders: Democrats Considering $6 Trillion Spending Package," *The Hill*, June 17, 2021.

Page 76 **the benefit of the doubt**: "'This Week' Transcript 6-27-21: Sen. Joe Manchin, Sen. Rob Portman, Keith Ellison & Mayor Charles Burkett," *ABC News*, June 27, 2021.

CHAPTER 9: EXIT, VOICE, AND LOYALTY

Page 79 **we were meeting**: We also worked closely with Sam Krasnow and Arjun Krishnaswami from NRDC.

Page 80 **at least 10 percent higher**: US Energy Information Administration, "Electric Power Sector CO_2 Emissions Drop as Generation Mix Shifts from Coal to Natural Gas," June 9, 2021.

Page 80 **published a report**: John Romankiewicz, Cara Bottorff, and Leah Stokes, *The Dirty Truth About Utility Climate Pledges* (Sierra Club, January 2021).

Page 82 **Biden's new climate target**: "'We Lost Time' on Action Against Climate Change in the Last Four Years," *MSNBC*, April 22, 2021.

Page 83 **Hope for Homes**: This program was first proposed in 2009 and called Home Star.

CHAPTER 10: PUMPS

Page 90 **her colleague**: Jane Flegal was Sonia's close colleague and collaborator in the Climate Policy Office. She worked on both the bipartisan infrastructure law and the climate package.

Page 92 **focus on getting the reconciliation bill done**: Allison Pecorin, Molly Nagle, and Trish Turner, "Biden-Capito Infrastructure Negotiations Have Ended," *ABC News*, June 8, 2021.

Page 93 **met regularly with the president**: Burgess Everett and Marianne Levine, "The Power of 10: Inside the 'Unlikely Partnership' That Sealed an Infrastructure Win," *Politico*, August 10, 2021.

CHAPTER 11: NO CLIMATE, NO DEAL

Page 96 **given an interview**: Zack Colman, "Biden Climate Adviser: Infrastructure Plan Could Omit Some Climate Proposals," *Politico*, June 8, 2021.

Page 96 **likely to be excluded**: David Morgan and Susan Cornwell, "U.S. Senators Push for Infrastructure Plan That Avoids Tax Hikes," Reuters, June 9, 2021.

Page 96 **wouldn't support an infrastructure deal**: Coral Davenport, Lisa Friedman, and Emily Cochrane, "Collapse of Infrastructure Talks Puts Climate Action at Risk," *New York Times*, June 11, 2021.

Page 96 **our rallying cry**: Jamal Raad, who led Evergreen at the time, was particularly gifted at guiding our communications strategy.

Page 97 **on the record**: Evergreen Action, "No Climate, No Deal: Action Hub," October 7, 2021.

Page 97 **lack of public mobilization**: David Roberts, "Why Did the Climate Bill Fail?," *Grist*, July 27, 2010.

Page 98 **require hundreds of billions**: Dan Esposito, *Studies Agree 80 Percent Clean Electricity by 2030 Would Save Lives and Create Jobs at Minimal Cost* (Energy Innovation, September 2021).

CHAPTER 12: INFRASTRUCTURE WEEK

Page 102 **both bills were equal priorities**: The White House, "Statement by President Joe Biden on the Bipartisan Infrastructure Framework," June 26, 2021.

Page 104 **new transmission had fallen by 80 percent**: Nathan Shreve, Zachary Zimmerman, and Rob Gramlich, *Fewer New Miles: The U.S. Transmission Grid in the 2020s* (Americans for a Clean Energy Grid, July 30, 2024).

Page 105 **direct 40 percent of the benefits**: The White House, "Justice 40: A Whole-of-Government Initiative."

Page 106 **Working with other activists**: Chispa and Mom's Clean Air Force supported these electric school bus campaigns.

CHAPTER 13: $3.5 TRILLION

Page 108 **continued to show disinterest**: Sahil Kapur, "Manchin Open to Biden's 'Human Infrastructure' Plans and Undoing Some Trump Tax Cuts," *NBC News*, June 22, 2021.

Page 109 **His wife also worked**: Meghan Taira was the legislative director for Senator Schumer at the time and played a central role in negotiating all legislation.

Page 112 **story about the deal**: Tony Romm, Seung Min Kim, and Jeff Stein, "Senate Democrats Reach $3.5 Trillion Budget Deal," *Washington Post*, July 13, 2021.

CHAPTER 14: THE WAITING ROOM

Page 113 **officially in the budget deal**: Dana Farrington and Barbara Spunt, "As Democrats Look to Cut Their Spending Bill, Here's a Reminder of Where They Started," *NPR*, October 12, 2021.

Page 115 **called on Congress**: Gina McCarthy and Anita Dunn, "White House Memorandum: Pushing President Biden's Full, Popular, and Bipartisan Build Back Better Agenda Forward," June 29, 2021.

Page 116 **new EV charging investments**: White House, "FACT SHEET: Historic Bipartisan Infrastructure Deal," July 28, 2021.

Page 116 **electric by 2030**: Jack Ewing, "President Biden Sets a Goal of 50 Percent Electric Vehicle Sales by 2030," *New York Times*, August 5, 2021.

Page 116 **just one in 16 were**: Joel Jaeger, "These Countries Are Adopting Electric Vehicles the Fastest," World Resources Institute, September 14, 2023.

Page 116 **Pollution had grown**: Clara Camarasa et al., "A Global Comparison of Building Decarbonization Scenarios by 2050 Towards 1.5–2°C Targets," *Nature Communications* 13 (2022): 3077.

Page 117 **not a substitute for climate action**: Leah C. Stokes, "The Infrastructure Bill Won't Cut It on Climate," *The Atlantic*, July 14, 2021.

CHAPTER 15: A ONE-PAGE AGREEMENT

Page 119 **to get Manchin to guarantee**: Joe Manchin, *Dead Center* (St. Martin's, 2025), 169.

Page 119 **one-page document**: Burgess Everett, "Manchin Proposed $1.5T Top-Line Number to Schumer This Summer," *Politico*, September 30, 2021.

Page 119 **burned him on that bill**: Manu Raju, Ted Barrett, Phil Mattingly, and Lauren Fox, "How Democrats Miscalculated Manchin and Later Won Him Back," CNN, March 7, 2021.

Page 120 **"Don't tell anybody"**: Manchin, *Dead Center*, 169.

Page 120 **send the document to Speaker Pelosi**: Franklin Foer, *The Last Politician: Inside Joe Biden's White House and the Struggle for America's Future* (Penguin, 2023), 220.

CHAPTER 17: THE NICU

Page 127 **scheduled for a vote in the Senate**: "Senate Takes Up $1 Trillion Bipartisan Infrastructure Deal, Signaling a Breakthrough for a Top Biden Priority," *New York Times*, July 28, 2021.

Page 127 **pressure from Speaker Pelosi**: Daniella Diaz and Clare Foran, "Pelosi: House Won't Take Up the Bipartisan Bill Until Senate Votes on Reconciliation," *CNN*, June 24, 2021.

CHAPTER 18: 45 PERCENT

Page 129 **would increase the deficit**: Alan Rappaport and Emily Cochrane, "Infrastructure Bill Would Add $256 Billion to Deficit, Analysis Finds," *New York Times*, August 7, 2021.

Page 129 **railed against the reconciliation bill**: Emily Cochrane, "Senate Passes $1 Trillion Infrastructure Bill, Handing Biden a Bipartisan Win," *New York Times*, August 11, 2021.

Page 131 **received the infrastructure bill poorly**: Li Zhou, "'Try Us': House Progressives Finally Flex Their Power," *Vox*, September 30, 2021.

Page 132 **hanging out with some of the utility**: Benjamin Storrow, Scott Waldman, and Lesley Clark, "Here's What Manchin Told the Biden Team on Climate," *E&E News*, October 18, 2021.

CHAPTER 19: TESTIMONY

Page 137 **he did not hold back**: "Meet the Press—September 12, 2021," *NBC News*, September 12, 2021.

Page 137 **also worried about reliability**: Daniella Diaz, Devan Cole, and Chandelis Duster, "Manchin: I Will Not Vote for $3.5 Trillion Bill," *CNN*, September 12, 2021.

Page 139 **I urged them to end this gridlock**: Leah C. Stokes, "Statement on 'Examining the Economic Benefits of Electrifying America's Homes and Buildings,'" Joint Economic Committee, September 22, 2021.

CHAPTER 20: BEGINNINGS AND ENDINGS

Page 141 **meeting with him and a group of utility executives**: Storrow et al., "Here's What Manchin told the Biden Team."

Page 143 **denied trying to shrink the package**: Coral Davenport, "Kyrsten Sinema Wants to Cut $100 Billion in Proposed Climate Funds, Sources Say," *New York Times*, October 8, 2021.

Page 143 **criticized for pushing cuts**: Maanvi Singh, "Climate Advocates Who Backed Sinema Exasperated by Blocking of Biden Bill," *Guardian*, October 22, 2021.

Page 147 **staffer from Senator Smith's office**: Pete Wycoff, Senator Smith's policy advisor for energy and environment, was essential to CEPP's policy development.

Page 147 **couldn't meet our climate goals**: Coral Davenport, "Key to Biden's Climate Agenda Likely to Be Cut Because of Manchin Opposition," *New York Times*, October 16, 2021.

CHAPTER 21: THE RACE TO SAVE TONS

Page 148 **modeling from a number of groups**: Dan Esposito, *Studies Agree 80 Percent Clean Electricity by 2030 Would Save Lives and Create Jobs at Minimal Cost* (Energy Innovation, September 2021).

Page 149 **to the billionaire Bill Gates**: Zack Colman, Josh Siegel, and Kelsey Tamborrino, "The Manchin Pressure Campaign: CEOs, Labor Bosses and Bill Gates," *Politico*, July 29, 2022.

Page 150 **Energy Department Staffers**: Jeremiah Baumann led this initiative at the DOE.

Page 152 **the country was way behind**: Daniel Klein, "CO_2 Emission Trends for the US and Electric Power Sector," *The Electricity Journal* 29, no. 8 (2016): 33–47.

Page 154 **different understanding**: Manchin, *Dead Center*, 174.

CHAPTER 23: BREAKDOWN

Page 158 **add several social policies**: Kelsey Snell, "The House Passes a $2 Trillion Spending Bill, but Braces for Changes in the Senate," *NPR*, November 19, 2021.

Page 159 **close to fixing it in 2018**: Michael Grunwald, "Why Trump Can't Kill the Electric Car," *Politico Magazine*, December 13, 2018.

Page 159 **overly reliant on batteries**: Hannah Northey, "Manchin on EV Deal: 'Ford Has Serious Questions to Answer,'" *E&E News*, February 15, 2023.

Page 159 **showed up at their house**: Ryan Grim, *The Squad: AOC and the Hope of a Political Revolution* (Henry Holt, 2023).

Page 159 **chief of staff**: Lance West served as Senator Manchin's chief of staff.

Page 160 **killing the entire bill**: Lisa Friedman and Coral Davenport, "Manchin Rejects Landmark Legislation, Putting Biden's Climate Goals at Risk," *New York Times*, December 19, 2021.

Page 161 **most destructive fire**: "Looking Back at Colorado's Marshall Fire," NOAA, January 8, 2024.

CHAPTER 24: DUNKELFLAUTE

Page 165 **electricity prices jumped 8 percent**: Ivan Penn, "Get Ready for Another Energy Price Spike: High Electric Bills," *New York Times*, May 3, 2022.

Page 165 **rose almost 10 percent**: AAA, "Global Tensions Push Oil Prices Higher, and Pump Prices Follow," January 31, 2022.

Page 165 **we were seeing momentum**: Jamal Raad, who led Evergreen at the time, drove this strategy.

Page 166 **joined a daily call**: Lori Lodes, the executive director of Climate Power, organized and ran these daily calls.

Page 168 **book on the topic**: Rowena Bennett, *Your Baby's Bottle-Feeding Aversion: Reasons and Solutions* (2017).

Page 169 **Manchin told reporters**: Alan Fram, "Manchin, Key Dem, Says Build Back Better Bill Is 'Dead,'" AP, February 1, 2022.

Page 169 **had to go through committee**: "Jake Tapper Asks Manchin About Future of Biden's Bill," *CNN*, February 6, 2022.

CHAPTER 25: REBIRTH

Page 174 **he was open**: Maxine Joselow et al., "Sen. Manchin Launches New Push for 'All of the Above' Energy Bill," *Washington Post*, March 25, 2022.

CHAPTER 26: THE DEFENSE PRODUCTION ACT

Page 175 **planning to wean themselves off**: Leila Fadel and Rob Schmitz, "European Union Leaders Agree to Ban 90% of Russian Oil by the End of 2022," *NPR*, May 31, 2022.

Page 175 **called on the federal government**: Bill McKibben, "Heat Pumps for Peace and Freedom," *The Crucial Years*, Substack, February 27, 2022.

Page 175 **put out retro-style wartime posters**: Sarah Lazarovic was the brilliant person behind these posters.

Page 176 **plan to reduce EU dependence**: International Energy Agency, *A 10-Point Plan to Reduce the European Union's Reliance on Russian Natural Gas* (2022).

Page 176 **"Electrify for Peace" plan**: Rachael Grace, Stephen Pantano, Leah Stokes, Ari Matusiak, and Sam Calisch, *Electrify for Peace Policy Plan* (Rewiring America, March 9, 2022).

Page 176 **oil prices jumped another 18 percent**: Maciej Kolaczkowski, "How Does the War in Ukraine Affect Oil Prices?," World Economic Forum, March 4, 2022.

Page 177 **calling on Schumer to act**: People's Action, Indivisible, Working Families Party, and Center for Popular Democracy, "Organizational Letter to Schumer Re Budget Reconciliation," March 14, 2022.

Page 179 **"Do you want to go ziplining?"**: Foer, *The Last Politician.*

Page 179 **had to keep listening**: Phil Mattingly, "'Keep Talking': How a Simple Philosophy Helped Lead to a White House Celebration," *CNN*, September 13, 2022.

Page 180 **named heat pumps**: White House, "Joint Statement Between the United States and the European Commission on European Energy Security," March 25, 2022.

CHAPTER 27: SOLAR INTERFERENCE

Page 182 **80 percent of solar imports overnight**: Matthew Daly, "A Commerce Department Investigation Could Imperil the Solar Industry, Advocates Say," *Los Angeles Times*, March 29, 2022.

Page 182 **over a third of new capacity**: American Clean Power, *2022 Annual Market Report* (May 22, 2023).

Page 183 **bit half a dozen people**: Dareh Gregorian and Kyle Stewart, "Police Nab Fox on Capitol Hill After Attack on Democratic Congressman, Other 'Encounters,'" *NBC News*, April 5, 2022.

Page 183 **he wanted to get the bill done**: Adam Cancryn and Eugene Daniels, "The New White House Rule: Do Not Talk About Joe Manchin," *Politico*, April 11, 2022.

Page 183 **a large network**: The Green New Deal Network was the lead organizer of this event.

Page 184 **targeting coal plants**: Mike Tony, "Activists Blockade Grant Town Coal-Fired Plant to Protest Manchin Blocking Climate Legislation While Making Millions from Coal," *Logan Banner*, April 20, 2022.

CHAPTER 28: MADE IN AMERICA

Page 187 **just 3,000 were sold**: John O'Dell, "In Hydrogen vs. Electric Cars Comparison, Who Wins?," Edmunds.com, March 25, 2024.

Page 188 **blew through their caps**: David Shepardson, "GM Sold 200,000 Electric Vehicles in U.S. by 2018, Triggering Tax-Credit Phaseout: Source," Reuters, January 2, 2019.

Page 189 **building new facilities**: Eric Wolff, "Manchin, Stabenow push $8B Tax Credit to Bring Factories for Former Coal Sites," *Politico*, March 3, 2021.

Page 189 **gave companies a bonus**: Department of Energy, "Federal Tax Credits for Solar Manufacturers," 2022.

Page 190 **consider West Virginia**: Anne Fischer, "Form Energy to Set Up an Iron-Air Battery Manufacturing Facility in West Virginia," *pv magazine*, December 23, 2022.

Page 190 **wanted to spend less**: Maxine Joselow and Tony Romm, "Democrats Race to Clinch Deal on Climate, Energy with Manchin," *Washington Post*, July 7, 2022.

Page 191 **longer transition increased spending**: Joint Committee on Taxation, "Estimation of Budget Effects of the Revenue Provisions of Title I," August 9, 2022.

Page 192 **offered fewer public lands**: Sam Zeno, "The Biden Administration Has Reached Conservation Records in 2023," Center for American Progress, December 20, 2023.

CHAPTER 29: DECLARATION

Page 193 **using the Defense Production Act to build more solar panels**: White House, "Memorandum on Presidential Determination Pursuant to Section 303 of the Defense Production Act of 1950, as amended," June 6, 2022.

Page 193 **just used the Defense Production Act**: White House, "Memorandum on the Delegation of Authority Under the Defense Production Act to Ensure an Adequate Supply of Infant Formula," May 18, 2022.

Page 196 **$235 million for the effort**: California Energy Commission, "California Schools Healthy Air, Plumbing, and Efficiency Program."

Page 196 **limit the federal government's authority**: Adam Liptak, "Supreme Court Strips Federal Government of Crucial Tool to Control Pollution," *New York Times*, June 30, 2022.

CHAPTER 30: GETTING TO YES

Page 197 **about to lapse**: Dena Bunis, "Expanded ACA Subsidies Extended Through 2025," *AARP*, August 15, 2022.

Page 198 **activists were delaying**: Carlos Anchondo, "Mountain Valley Pipeline Delayed: 4 Questions Answered," *E&E News*, October 19, 2023.

Page 199 **deliver a 45 percent reduction**: Senate Majority Leader Schumer, "Dear Colleague" (letter), August 25, 2021.

Page 199 **secure $7.5 billion**: ZETA, "Electric Vehicle Provisions in the Infrastructure Investment and Jobs Act," February 1, 2023.

Page 199 **$30 billion**: Emily Cochrane and Jonathan Weisman, "House Narrowly Passes Biden's Social Safety Net and Climate Bill," *New York Times*, November 19, 2021.

CHAPTER 31: THINGS FALL APART

Page 203 **hottest in world history**: Jeff Masters, "NASA: June 2022 Tied as Earth's Warmest June on Record," *Yale Climate Connections*, July 14, 2022.

Page 203 **limited the authority**: Emma Newburger and Dan Mangan, "Supreme Court Limits EPA Authority to Set Climate Standards for Power Plants," *CNBC*, June 30, 2022.

Page 204 **largest increase in over 40 years**: Lucy Bayly and Alicia Wallace, "US Inflation Hit 40-Year High in June, Driven by Record Gas Prices," *CNN*, July 13, 2022.

Page 205 **broken down for good**: Tony Romm and Jeff Stein, "Manchin Says He Won't Support New Climate Spending or Tax Hikes on Wealthy," *Washington Post*, July 15, 2022.

Page 205 **claimed he had not walked away**: Emily Cochrane and Lisa Friedman, "Manchin Pulls Plug on Climate and Tax Talks, Shrinking Domestic Plan," *New York Times*, July 14, 2022.

Page 206 **more campaign contributions**: Coral Davenport, "This Powerful Democrat Linked to Fossil Fuels Will Craft the U.S. Climate Plan," *New York Times*, September 19, 2021.

Page 206 **half a million dollars each year**: Christopher Flavelle and Julie Tate, "How Joe Manchin Aided Coal, and Earned Millions," *New York Times*, March 27, 2022.

CHAPTER 32: THE LONGEST TWO WEEKS

Page 208 **he hadn't walked away**: Brad McElhinney, "Manchin Says Inflation Concerns Led to Impasse on Latest Build Back Better Talks," *MetroNews*, July 15, 2022.

Page 209 **her White House colleagues**: Maggie Thomas, who served as special assistant to the president for climate and senior advisor for strategic planning at the Climate Policy Office, played a key role in drafting the climate emergency declaration.

Page 210 **screwed over West Virginia**: Jael Holzman, "Manchin Veto of Climate Bill Deals Setback to Coal Miners," *E&E News*, July 27, 2022.

Page 210 **ready to pass the CHIPS bill**: Lindsey McPherson and Laura Weiss, "Senate GOP Pitches Options for Stalled Competition Bill," *Roll Call*, July 12, 2022.

Page 210 **"Congress should act"**: Joseph Zeballos-Roig, "Manchin Says He Wants Congress to Pass Climate Programs After He Tanked Passage of Climate Programs," *Business Insider*, July 19, 2022.

Page 211 **said the word *emergency* five times**: White House, "Remarks by President Biden on Actions to Tackle the Climate Crisis," July 20, 2022.

Page 211 ***Maybe* by Kobi Yamada**: Kobi Yamada, *Maybe*, ill. Gabriella Barouch (Compendium, 2019).

Page 212 **"Never Say Die" caucus**: Akshat Rathi and Jennifer A. Dlouhy, "Bill Gates and the Secret Push to Save Biden's Climate Bill," *Bloomberg*, August 16, 2022.

Page 212 **Prove every critic wrong**: Tony Romm and Jeff Stein, "The Two-Week Scramble That Saved Democrats' Climate Agenda," *Washington Post*, July 30, 2022.

Page 214 **he supported the bill**: Senate Committee on Energy and Natural Resources, "Manchin Applauds Signing of CHIPS and Science Act," August 9, 2022.

CHAPTER 33: "THE INFLATION REDUCTION ACT"

Page 217 **chief of staff**: Mike Lynch served as Senator Schumer's chief of staff.

Page 219 **Republicans were now pushing**: Megan Mineiro, "This Week: House Begins Busy Sprint to August Recess, Senate Tees Up 'Chips-Plus' Bill," *Roll Call*, July 25, 2022.

Page 220 **this has got to be it**: Foer, *The Last Politician*, 351.

Page 221 **scheduled to arrive later**: Foer, *The Last Politician*, 352.

Page 221 **"Are you still upset?"**: Alexander Bolton, "Inside the Secret Manchin-Schumer Deal: Dems Shocked, GOP Feels Betrayed," *The Hill*, July 29, 2022.

Page 221 **"As long as we finish in August"**: Romm and Stein, "The Two-Week Scramble."

CHAPTER 34: THE FIGHT OF ENACTMENT

Page 225 **four-page summary**: Senate Democrats, "Summary of the Energy Security and Climate Change Investments in the Inflation Reduction Act of 2022," July 27, 2022.

Page 226 **also had to be offered**: Jake Bittle, "The Inflation Reduction Act Promises Thousands of New Oil Leases. Drillers Might Not Want Them," *Grist*, August 9, 2022.

Page 226 **just 12 percent of areas**: Sarah Hart-Curran, "Potential Impacts of the IRA's Onshore Energy Leasing Provisions," Harvard Law School Environmental & Energy Law Program, March 24, 2023.

Page 226 **permitting reform bill**: Nina Lakhani, "Schumer and Manchin's 'Dirty Side Deal' to Fast-Track Pipelines Faces Backlash," *Guardian*, September 22, 2022.

Page 226 **20 times**: Megan Mahajan, Olivia Ashmoore, Jeffrey Rissman, Robbie Orvis, and Anand Gopal, *Updated Inflation Reduction Act Modeling Using the Energy Policy Simulator* (Energy Innovation, August 23, 2022).

Page 226 **pushing the bill**: "Manchin Touts 'Red, White and Blue' Reconciliation Deal Despite Recession Woes: 'Tremendous Opportunity,'" *Fox News*, July 31, 2022.

Page 228 **Department of Energy released**: Department of Energy, "DOE Projects Monumental Emissions Reduction From Inflation Reduction Act," August 18, 2022.

Page 229 **tens of thousands of dollars**: Emily Atkin, "Beware of This Fake Pro-Climate Group." *Heated*, Substack, August 5, 2022.

CHAPTER 35: THE BYRD BATH

Page 232 **proved particularly helpful**: Bill Dauster and Mike Jones were instrumental in planning Democrats' arguments to the parliamentarian.

Page 235 **Sinema remained uncommitted**: Nick Sobczyk and Jeremy Dillon, "Sinema Strikes Deal on Reconciliation; Schumer Sets Vote," *E&E News*, August 5, 2022.

Page 235 **intense conflict**: Maxine Joselow, "Inside the Two-Day Scramble to Add Drought Funding to the Climate Law," *Washington Post*, September 23, 2022.

CHAPTER 36: 27 HOURS

Page 238 **Manchin's opposition**: Grace Segers, "Senate Passes Biden's $1.9 trillion COVID Relief Bill After 'Vote-a-Rama,'" *CBS News*, March 7, 2021.

Page 238 **criticize the bill**: Merdie Nzanga, "Bernie Sanders Criticizes Inflation Reduction Act, Says It Will Have 'Minimal Impact,'" *USA Today*, August 7, 2022.

Page 238 **"a tax increase"**: "Graham Calls Out Manchin-Schumer Bill for Not Addressing Inflation," posted August 6, 2022, by Budget GOP, YouTube, 20 min., 19 sec.

CHAPTER 37: OF EARTH

Page 245 **call her boss**: Ali Zaidi was the deputy national climate advisor to President Biden.

CHAPTER 38: CELEBRATION

Page 251 **proud to introduce**: IBEW, "Celebrating Progress for Workers, Boston Member Hits 'Home Run' at White House," November 3, 2022.

CHAPTER 40: HAPPY BIRTHDAY

Page 260 **My friends**: Rachel Fakhry and Sam Krasnow spearheaded the 45V campaign.

Page 260 **$33 billion in total requests**: Jean Chemnick, "EPA to Announce Billions in Climate Grants for States, Cities," *E&E News*, July 22, 2023.

Page 261 **projected to cost almost $73 billion**: Nicholas E. Buffie, "The Section 45X Advanced Manufacturing Production Credit," Congressional Research Service, November 7, 2024.

Page 261 **almost no bidders**: "US Holds First of Several Oil and Gas Auctions as COP28 Gets Underway," Reuters, November 28, 2023.

Page 262 **Manchin had soured**: Senator Joe Manchin (@Sen_JoeManchin), "Statement on the One-Year Anniversary of the Signing of the Inflation Reduction Act," Twitter, August 16, 2023.

Page 263 **end up supplying 96 percent**: McCarthy, "Chart: 96 Percent of New US Power Capacity."

Page 263 **one new gas plant**: US Energy Information Administration, "Electric Generators Plan More Natural Gas-Fired Capacity After Few Additions in 2024," June 11, 2025.

Page 263 **upward of seven years**: Jared Anderson, "US Gas-Fired Turbine Wait Times as Much as Seven Years; Costs Up Sharply," *S&P Global*, May 20, 2025.

Page 263 **12,000 across the country**: Francesca Paris and Eve Washington, "See How E.V. Road Trips Went from Impossible to Easy," *New York Times*, September 29, 2025.

Page 263 **$129 million in funding**: Environmental Protection Agency, "EPA Announces More Than $129 Million to Cuyahoga County for Community-Driven Solutions to Cut Climate Pollution," press release, July 22, 2024.

Page 264 **too old**: Meredith Deliso, "Overwhelming Majority of Americans Think Biden Is Too Old for Another Term: POLL," *ABC News*, February 11, 2024.

EPILOGUE

Page 269 **reported that Trump**: Josh Dawsey and Maxine Joselow, "What Trump Promised Oil CEOs as He Asked Them to Steer $1 Billion to His Campaign," *Washington Post*, May 9, 2024.

Page 270 **four in five incumbent governments**: David Rising, Jill Lawless, and Nicolas Ricciardi, "The 'Super Year' of Elections Has Been Super Bad for Incumbents as Voters Punish Them in Droves," AP, November 17, 2024.

Page 271 **harassing organizations**: Jean Chemnick, "Trump Officials Escalate Fight to Take Back $20B in Climate Money," *Politico*, March 8, 2025.

Page 271 **3,500 percent duties on solar panels**: João da Silva, "US Sets Tariffs of Up to 3,521% on South East Asia Solar Panels," *BBC News*, April 22, 2025.

Page 272 **ramped up 500 percent**: Solar Energy Industries Association, *Solar Market Insight Report Q2 2025* (June 9, 2025).

Page 272 **$170 billion of private capital**: Clean Economy Tracker, 2025, https://cleaneconomytracker.org/.

Page 272 **still had fossil fuel interests**: Dorothy Slater, "Clean-Power Group Supports Dirty Energy," *The American Prospect*, April 28, 2023.

Page 272 **impossible red tape**: Scott Streater, "Interior's New Solar, Wind Policy Sparks Concerns of 'Shadow Ban,'" *E&E News*, July 18, 2025.

Page 272 **"disgusting abomination"**: Joey Garrison, "Elon Musk Slams President Trump's Big Tax and Policy Bill As a 'Disgusting Abomination,'" *USA Today*, June 3, 2025.

Page 273 **undercut the Republican senators' attempts**: Diana DiGangi, "Trump Seeks Tighter Restrictions on Wind and Solar with Executive Order," *UtilityDive*, July 9, 2025.

Page 274 **437,000**: Dan McCarthy, "Chart: EV Sales Just Hit a Record in the US, but a Cliff Looms," *Canary Media*, October 17, 2025.

Page 274 **70 percent year over year**: Kelly Pickerel, "Q4 2025 Expected to Be Record-Breaking Quarter for Residential Battery Installs," *Solar Power World*, December 16, 2025.

Page 274 **one oil executive**: Jamie Smyth and Kristina Shevory, "Opec Oil 'Price War' Will Halt Shale Boom, Say US Producers," *Financial Times*, August 14, 2025.

Page 274 **supply chain for gas power plants was broken**: Jason Plautz, "Want to Build a Gas Plant? Get in Line," *E&E News*, April 22, 2025.

INDEX